SMALL SCALE FOUNDRIES FOR DEVELOPING COUNTRIES

A GUIDE TO PROCESS SELECTION

J.D. Harper

An Intermediate Technology Publication

Acknowledgements

The printing of this publication has been made possible by a grant from the Overseas Development Administration. The Intermediate Technology Development Group gratefully acknowledges their generosity.

Written by Geoffrey Lamb (Consultants) Ltd of Birmingham, U.K. for Intermediate Technology Industrial Services, Rugby, U.K. Published by Intermediate Technology Publications Ltd, 103/105 Southampton Row, London WC1B 4HH, U.K.

ISBN 0 903031 78 7

Printed by Antony Rowe Ltd, Chippenham, Wiltshire (U.K.)

CONTENTS

List of illustrations

CHAPTER 1

INTRODUCTION

The Role Of The Foundry

Castings in iron, brass, aluminium, or other metals are an essential part of most engineering products, and a foundry in which to make them is needed by any developing industrial society. Although the production of castings on a large scale is a sophisticated and capital intensive business, there can be a useful market for small scale foundries producing castings for building and domestic products, machinery parts, and spare parts for other equipment.

Some foundries are independent organisations producing castings for a number of customers. Other foundries are departments of larger concerns, which need their own source of castings for their products, or for the production of spare parts.

Foundries may specialise in one or more types or sizes of casting, or produce a wide range of types of product. Some castings are used for standard items such as pipe fittings, manhole covers and cooking stoves, whilst others have to be made individually for each application.

Many foundries have their own machining workshops to produce finished components or products, while others only make raw castings.

Foundry technology is changing rapidly, for small foundries as well as for large; casting production is becoming more of a science with modern techniques, and less of a traditional art. Nevertheless many valuable skills have to be learned and experience as well as theoretical knowledge is needed by the foundryman.

Some of the more commonly required types of casting which might be produced by small scale foundries are listed below.

Such a list can never be complete — one of the advantages of the foundry process is its flexibility, and the possibility of making all sorts and types of casting for a wide range of applications.

Cast Iron Castings	*Aluminium Castings*	*Brass & Bronze Castings*
Stoves	Levers & handles	Valves & taps for corrosive liquids
Pulleys	Fan and motor housings	
Manhole covers	Cooking pots and kitchen tools	Bearings and bushes
Pipe fittings		Boat parts and propellors
Pumps	Portable pump bodies	
Fire bars		Pump bodies and pump impellors
Brake drums	Pulleys	
Vehicle Spares	Irrigation pipe fittings	Ornamental and decorative castings
Bearing blocks	Light fittings	
Valves	Door furniture	Door furniture
Machinery spares	Machinery spare parts for all types of equipment	Machinery spare parts for all types of equipment

1

Object Of This Book

This book is not intended as a textbook of foundry practice. The purpose is rather to assist anyone about to start or to expand a small scale foundry to consider the various available processes, and to select the most appropriate for the circumstances.

An indication is given of the type of raw materials and equipment which will be needed, and the degree of training or skill which is likely to be required.

Foundry Processes

Foundry processes comprise the following elements:

1. Making a pattern, which is a model of the casting to be produced. This pattern is used many times in casting production.
2. Making a mould from the pattern. Usually the mould is made of sand and is only used once; but sometimes permanent metal moulds are used.
3. Melting the metal.
4. Pouring the metal into the mould.
5. Removing the solidified castings from the mould after cooling.
6. Cleaning and finishing the casting.

Despite these common elements, there is a great variety of processes available for each step. Different processes are suitable for different sizes and types of casting, for different metals, and for different degrees of mechanisation depending on the scale of the market and the number of each casting to be made. Some processes require costly raw materials or special purpose machinery, whilst others may require more labour or more skill than others.

Few foundries can make all possible types of casting although, according to the circumstances, some foundries may be more specialised than others.

Depending upon the range of products required, and upon the raw materials and capital available, it is possible to select the most appropriate processes for any foundry.

CHAPTER 2

PATTERN-MAKING

Almost all casting production starts with the manufacture of a pattern. Pattern design and construction has to be carefully considered in relation to the foundry process, as well as to the design of the finished casting. Much of the skill in foundry work is contained in the skill with which patterns are designed and made. Nevertheless, the basic principles are relatively simple, and intelligent craftsmen skilled in working with wood or metal can learn an adequate amount of elementary pattern-making with a minimum of training.

Existing Castings as Patterns
It is possible to use an existing casting as a pattern. This technique can sometimes be used when a replacement is needed for a damaged part, particularly when dimensional accuracy is not very important. However this method cannot readily be used with complicated designs, requiring the use of internal cores, as described below. Finished parts have often been machined, so that castings produced from them will not carry enough material thickness for their own machining operation.

Patterns have to be made larger than the castings to allow for one or two per cent contraction of the metal on solidification; a consequence of this is that the use of an old casting for a pattern will produce a new casting which is smaller than the original.

Sometimes these problems can be overcome by building up the casting locally with wood, wax, plastic or other material before it is used as a pattern. However, it is generally more satisfactory and sometimes quite essential to make a new pattern, especially if more than one or two castings are required.

Pattern Design
In designing a new pattern it is usually necessary to work from a scale drawing of the casting to be produced. The pattern designer must decide upon the mould joint plane, and upon which, if any, sections of the castings need to be formed by separate cores (internal cavities and sections which contain over-hangs which would otherwise prevent the mould from being separated from the pattern).

The pattern designer must calculate the dimensions in order to provide sufficient machining allowance, and to allow for metal contraction according to the metal or alloy to be used. The pattern must be tapered towards the mould joint to allow it to leave the sand cleanly in forming the mould. The pattern may have to include core prints, which are additional sections to form parts of the mould in which the cores are subsequently located. The pattern designer must also design the core boxes which are separate portions of pattern equipment used to make the cores.

The limitations of the foundry process have to be considered when designing patterns. Metal sections must not be too thin, nor must the pattern demand the moulding of unsupported thin pieces of sand. Sharp corners and re-entrant angles must be avoided.

The construction of the pattern must be related to the moulding process to be used, whether by hand or by machine, and according to the size of the casting being made. Patterns used for hand moulding or for the production of small quantities of castings may be made of wood, while patterns for production of castings by moulding machine are usually made of metal or plastic.

In order to simplify moulding, by whatever process, patterns for repetition production are often split and mounted on plates or boards at the mould joint level. Several small patterns may be mounted on the same plates. This technique requires accurate location between the two halves of the pattern and the mould, to avoid producing offset "cross jointed" castings. Mounting patterns for hand moulding can save considerable skill and effort on the part of the moulder. If moulding machines are used, it is essential to mount the patterns, usually on a steel or iron plate made to fit the moulding machine.

The quality of a casting can never be better than the quality of the pattern from which it is made. Good patterns are essential for good casting production.

Pattern Materials

Many patterns are made of wood. If only one or two castings are required, soft wood may be used. However if the pattern is to be used to make many castings, or to be stored for future requirements, the wood must be carefully selected. Close grained wood is needed, and the wood should be well seasoned to minimise the risk of warpage or distortion.

Wooden patterns must be strongly constructed, and be smooth and well finished. Unpainted wood may be used for some moulding processes, although special lacquer paints are available for patterns. Some kinds of paint stick to foundry sands and are not suitable.

Metal patterns for repetition castings may be made of aluminium, from bronze or from cast iron. Often they are themselves produced as castings from wooden patterns, subsequently hand finished or machined and polished to produce the working pattern. Sometimes duplicate patterns can most conveniently be made from epoxy resin plastic with fibreglass reinforcement. Using plaster moulds, high quality plastic patterns may be produced more easily (and with less skill) than by many other methods.

Good quality wooden patterns, used for hand moulding and stored with care, should last for the production of 50 to 100 moulds without major repair. Metal patterns used for machine moulding should last for many thousands of moulds. Aluminium is cheaper, but bronze lasts longer and is simpler to modify and repair by soldering.

Sometimes, when only one casting is required, a pattern is made from foamed polystyrene (as used for packing). This material can be cut with a sharp knife, a saw or a heated wire, and joined with glue or pins. Foamed polystyrene patterns are left in the mould and not removed. The molten metal when poured burns away the polystyrene (which is largely air) as it enters the mould cavity. The resulting castings are often rough but are suitable for some applications.

Pattern-Making Equipment

In order to make accurate patterns capable of producing high quality castings, it is necessary that adequate facilities are available. Equipment for making simple engineering drawings and the skills to prepare, read and interpret drawings are required.

Pattern-making tools include all conventional woodworking and metal working hand tools. Accurate measuring equipment is essential (rules, verniers, scales etc.) and a flat reference surface is necessary.

If power tools are available a disc sander, a planing machine, and a band saw will increase the range of patterns which can be made, reduce the labour content, and improve the accuracy of the work. A lathe is needed for many types of pattern.

For metal pattern-making a multi-speed drilling machine, an accurate lathe, and for some types of pattern a turret type milling machine are likely to be most useful.

The pattern shop must be well lit, and provided with adequate benches, vices, and tool and material storage space.

Parts of patterns must be located together accurately. It is also necessary to provide accurate location for coreboxes and for moulding boxes. For these purposes accurate bushes (sockets) and pins or dowels are needed. For the most accurate work these should be purchased from specialist manufacturers and made in hardened steel. However, for less critical castings, simple metal location pegs and bushes may be made in pattern shops. The use of wooden location pegs is not recommended.

A good supply of screws and bolts of different sizes is also necessary as is a good quality adhesive and a hard setting "filler" material.

Pattern Storage

If repeat casting orders are likely, patterns should be stored with care — especially since the pattern is often (depending upon the commercial arrangements) the property of the foundry's customer. Wooden and metal patterns should be kept on racks away from the foundry and the risks of accidental bruising or damage, fire hazards or damp.

In designing a foundry, adequate space should be provided for pattern storage.

Pipe bend casting

Top half mounted pattern

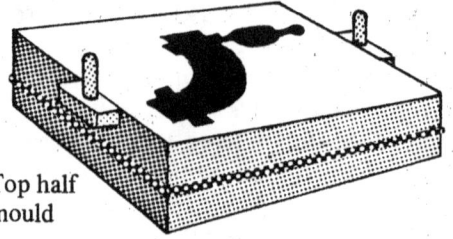

Top half mould

Bottom half mounted pattern

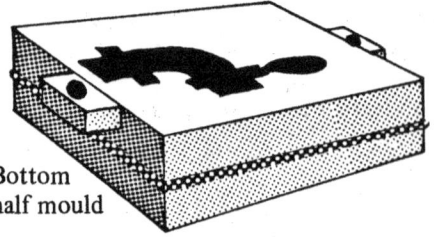

Bottom half mould

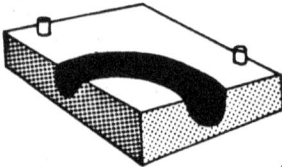 Corebox

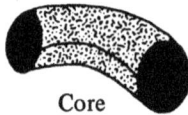

Core

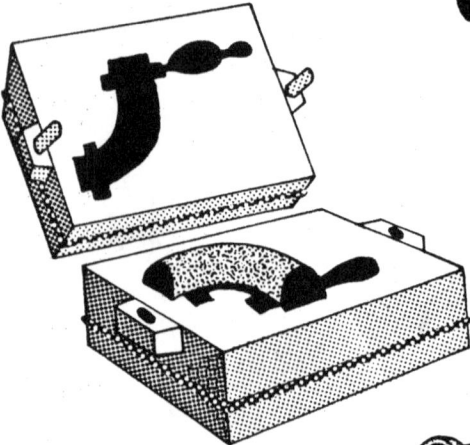

Assembling the mould

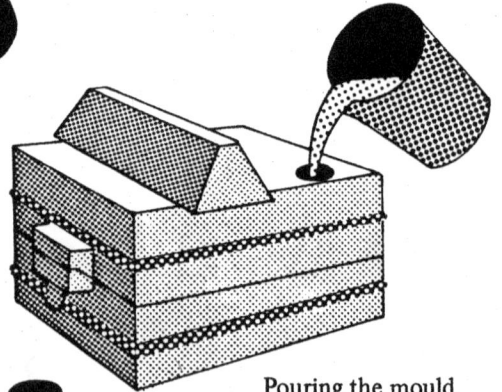

Pouring the mould

Casting ready for cleaning

Stages in Moulding a Cored Casting

6

CHAPTER 3

MOULDING

Choice of Moulding Method
The method of moulding to be used must be related to the type of castings to be produced and to the skills and equipment available in the foundry.

Small castings are usually produced in sand moulds, by hand if the quantities are not large, or on moulding machines for repetition work. Larger castings may also be made with moulding machines, although large machines are expensive. It will be necessary to handle large moulds with a crane, whether these are made by machine or by hand.

Running and feeding
The choice of the moulding method must also be related to the methods used to introduce the molten metal into the casting cavity through the runner system.

A typical runner system consists of a basin, formed in the top of the mould, to receive the metal as it is poured from the ladle. From this basin a vertical channel, called a downsprue, leads to the mould joint level, where horizontal channels known as runner bars lead to the casting cavity. The metal flows into the casting cavity from the runner bars through entry positions known as ingates.

As the metal in the casting cools and solidifies, it contracts. Unless more liquid metal is able to flow in to keep the cavity full, the casting will solidify with empty spaces or porosity. The additional liquid metal is provided by the use of feeders. Feeders are masses of metal, larger in section than the casting, joined to the ingate, and calculated to remain liquid until after the casting is completely solid.

Much of the skill of casting production lies in the way in which the runner and feeder systems are designed. The metal must flow freely into the thinnest sections of the casting, without scouring and washing away the sand. Slag and dirt should be prevented from entering the casting. Porosity and shrinkage must be avoided, by careful design of the solidification process, using feeders (and in some cases chills, which are metal inserts in the mould to accelerate cooling locally). At the same time the yield of the weight of casting to the weight of runner system has to be kept as high as possible to minimise melting costs.

Different metals, different casting designs, different moulding methods, and different pattern-making methods require different types of runner system design. Methods correct for one foundry may not be suitable for another foundry.

There are no fixed rules, the best methods being developed by experience and trial and error.

Some foundries rely upon moulders to design the runner system, whilst in other foundries this is the responsibility of the pattern designer or the manager.

Even in the small scale foundry, much time, effort and cost can be saved if experienced advice can be obtained in the field of runner system design.

Moulding Boxes
To produce most types of mould, prepared moulding sand is rammed around the pattern. Usually the pattern is set in a frame or moulding box. Moulding boxes may be made of iron, steel or wood. Moulding boxes must be accurately constructed, par-

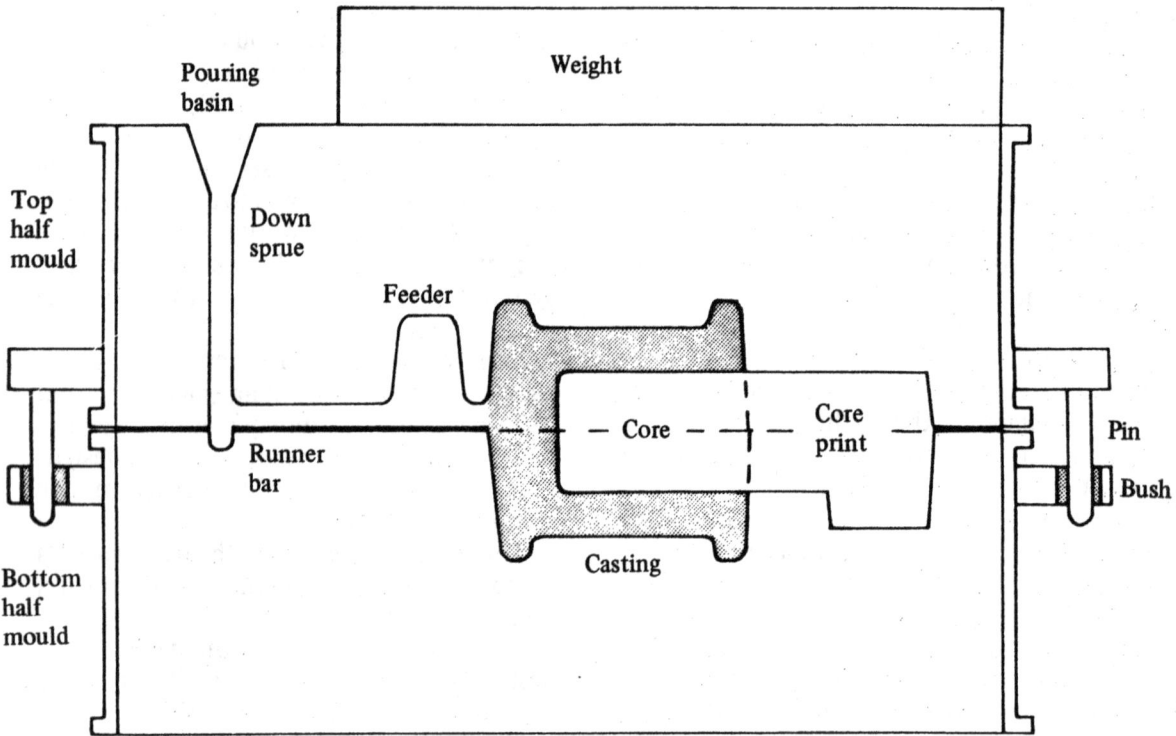

Parts of a Typical Mould (sectional view)

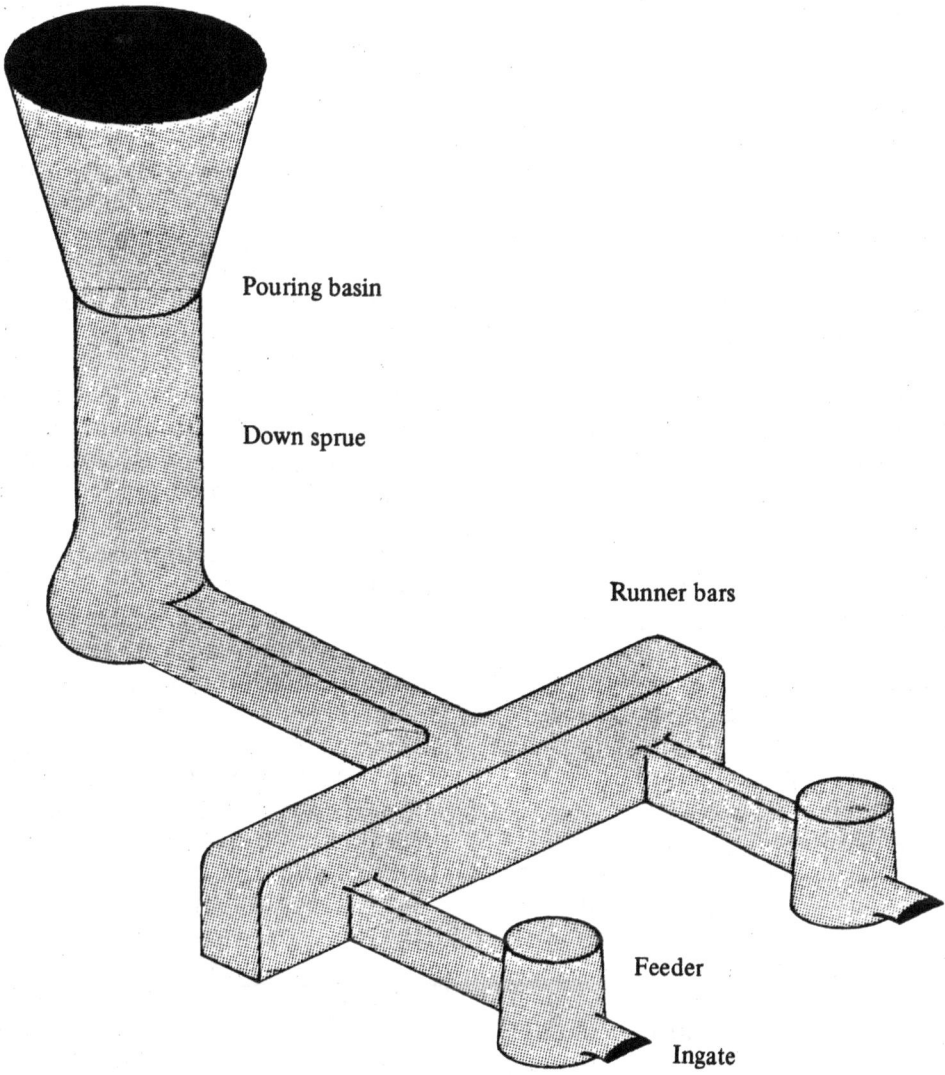

Pouring basin

Down sprue

Runner bars

Feeder

Ingate

A Runner System

ticularly in regard to the location of the two halves together. The production of large castings may require moulding boxes with reinforcement.

Reinforcing bars for moulding boxes may have to be welded or cut into different configurations for the production of different castings. Moulding box location should be ensured with accurate dowels and bushes, although for some less critical types of casting location strips and pegs can sometimes be accurate enough.

Some foundries do not use bottom half moulding boxes, embedding the lower half of the pattern directly into a sand floor. Except for very large castings, this method is not recommended since accuracy is difficult to maintain, and since it is not easy to ram sand round a lower half pattern. Very simple flat castings may be produced without moulding boxes at all, by "printing" the pattern into a prepared sand floor, and then pouring metal into the open cavity without the use of a top half mould. Not many castings are suitable for production by this ancient process.

Moulding box sizes have to be selected to suit the size of the casting to be made. One or more castings may be made in each mould — provided that they are to be cast in the same metal and poured together.

Small castings should be made in moulding boxes large enough to give at least 3 cm clearance around the edge of the pattern, and larger castings need more clearance.

Small moulds can sometimes be made in special moulding boxes which are hinged to open so that they can be removed from the mould after closing. If the sand is strong enough, and the weight of metal to be poured is not too great, the mould can stand on its own without support until the metal is poured and the casting is solid.

The advantage of this method is that only one moulding box is needed to produce a series of moulds. Moulding boxes of this type ("snap flasks", or "pop-off flasks") may be purchased, or made — if help is available for the design — in a pattern shop.

Permanent Moulding

Although most castings are made in sand moulds, which are only used once, it is possible to use permanent moulds made of metal for the production of a series of castings. The use of these permanent moulds or dies is more common for aluminium, zinc alloy, lead and brass castings than for other metals.

The metal mould, which may be constructed from steel or from high quality cast iron, must be produced accurately. A well equipped toolroom or machine shop is needed. The die should be designed by an experienced tool-or pattern-maker, ensuring that there is adequate taper, proper location, sufficient thickness of metal to withstand the heat of pouring, correct gating and feeding arrangements, methods for locating any cores, and arrangements for clamping the mould together. Permanent moulds may be coated with clay-silicate or graphite mixtures before use, and are preheated.

Many die-cast, permanently moulded parts, are produced on special purpose machines which inject the metal into the mould under high pressure. Pressure die-casting tooling and pressure die-casting machines are specialized and very expensive. The process is suitable for the mass production at low unit cost of very large numbers of components in zinc or aluminium.

For smaller scale production the gravity die-casting process may be used. The moulds are arranged to open and close and be clamped together manually or mechanically, as they are likely to be too hot and heavy for hand operation. Metal is poured manually and after solidification the casting is removed.

Gravity die-casting of aluminium, brass, lead or cast iron is a process which is particularly suitable for the production of simple casting shapes in large quantities. Although the cost of the metal moulds is high (higher than the cost of patterns for sand moulding) the fact that no sand, sand preparation, or sand-handling equipment is needed may

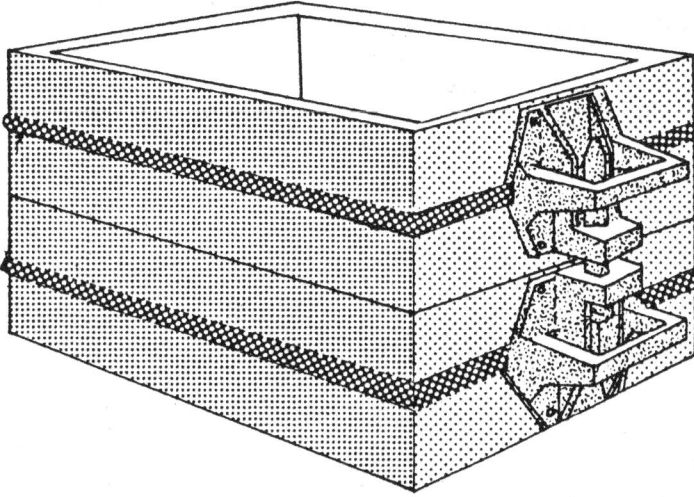

Pair of small moulding boxes with handles
pins and bushes

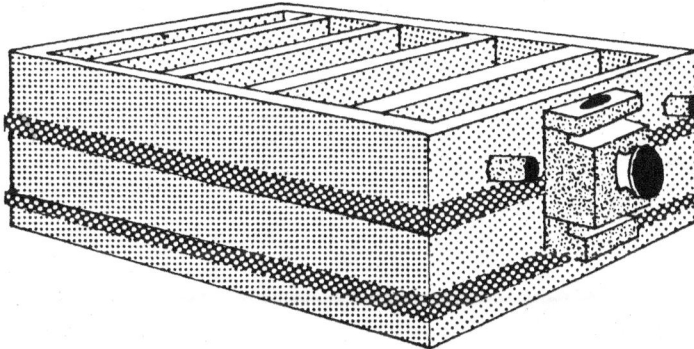

Half of a large moulding box with reinforcement
bars, location bushes, and trunnions for lifting
with a crane

Moulding Boxes

Strip and peg location system

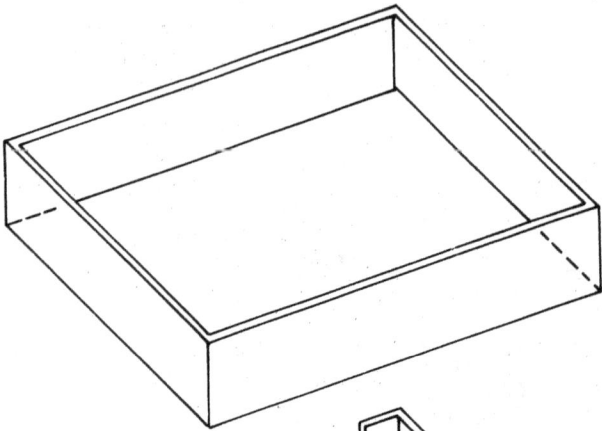

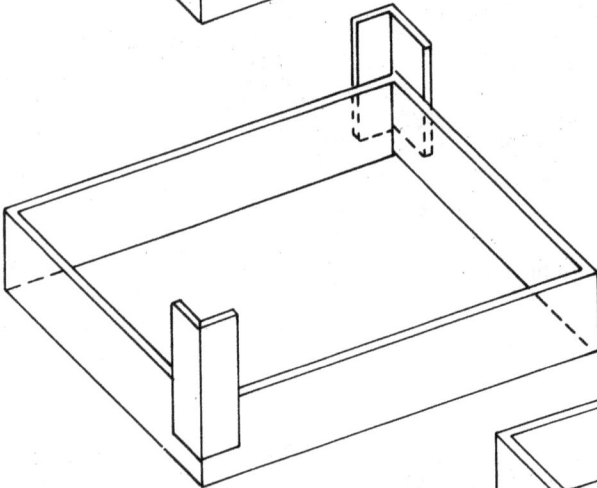

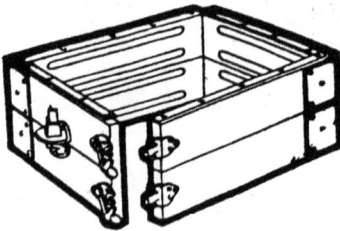

Corner location system

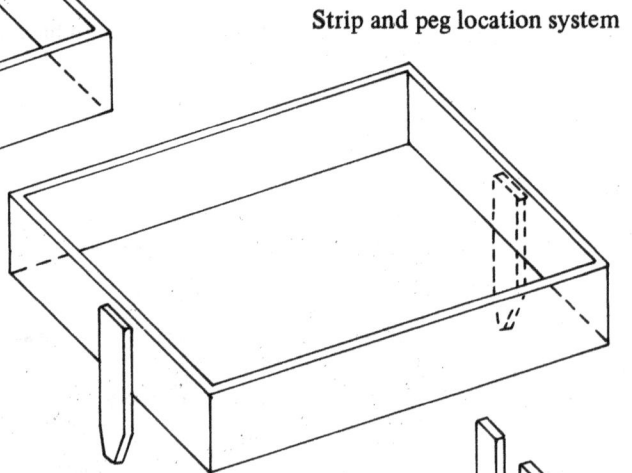

Snap — flask opened for removal

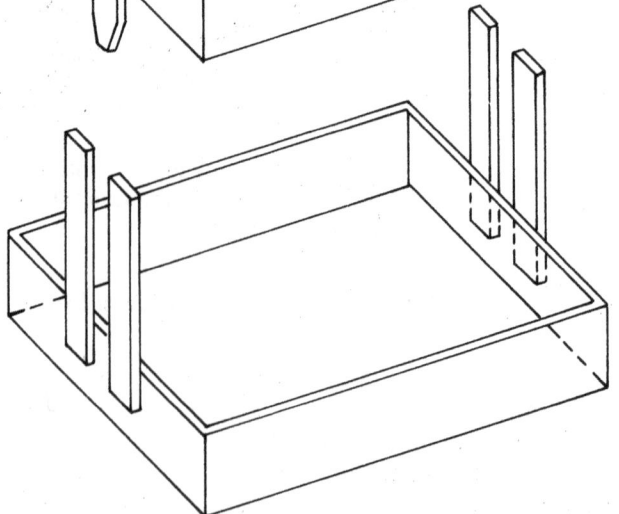

Alternative Types of Moulding Box

12

make the process more economical in some circumstances.

Centrifugal Casting

If molten metal is poured into a rapidly spinning tubular metal permanent mould a cylindrical casting can be made.

On a large scale, this process is used for cast iron and SG (spheroidal graphite) iron water-pipes; smaller centrifugally cast bronze tubes are used for making bearings, and cast iron cylinder liners for engines are also produced by this process.

The speed of rotation has to be related to the castings being made. For general castings, such as bronze bushes, about 200 rpm is normal, whilst for the highest quality cast iron cylinder liners speeds up to 900 rpm are used.

A centrifugal casting machine consists of a cast iron tubular mould, a motor and driving system for the spinning action, removable end plates and — usually — a water spray system to cool the outside of the mould. Machines may be purchased in a range of sizes; simple centrifugal casting machines can be made in a well equipped engineering workshop.

This process should not be confused with continuous casting in which metal flows slowly out of a furnace through a water-cooled collar, emerging as a rod or bar of solid metal. Continuous casting is not easy to control and the large amounts of material produced mean that it is a process unlikely to be of interest to small scale foundry enterprises.

Clay-Bonded Sand Moulds

Most castings are made in sand moulds. Not every sand is suitable for the production of castings. In order to produce a mould, the sand grains must be stuck or bonded together. The bond is usually provided by clay. Many sand deposits contain sufficient natural clay for this bond. Such "natural" sands with about 12% to 15% of clay are found in many parts of the world. It is possible by laboratory testing to determine whether a sand will be suitable for a foundry process, although it is usually better to carry out actual trials in a foundry.

Many foundries use sand which contains no natural clay, and add clay separately (usually between 5% and 10% by weight). This type of sand must be clean, and in particular free from mica, volcanic ash, crushed sea shell or coral. The grain size should be fairly uniform, the fineness determining the smoothness of the casting. The grain shape should be round, or sub-angular; wind blown sands or sands which have been washed as a by-product of mineral extraction are often suitable. Lake sands and river sands are frequently used, but sea beach sand is sometimes contaminated with shell or salt. Sand from some deposits may require washing and cleaning before use, or at very least sieving to remove lumps or foreign matter. Washed sand is likely to contain too much water for foundry use and may require drying — a simple sand dryer is not difficult to construct.

Clay and Moisture

Where separate clay additions are made, the best type of clay to use is bentonite. This material is available commercially for foundry, and also for oil-well use. Fireclay and other types of clay are also used in certain cases.

Sand bonded with clay must have the right amount of moisture in order to make good moulds. The water content depends upon the amount and type of clay present, varying from about 3% to about 7% by weight. The water content can be measured by accurately weighing a sample of sand before and after drying in an oven — other chemical and electrical methods can also be used in laboratories.

Mould Drying

Moulds made in clay-bonded sand, with a natural or synthetic clay, are suitable for the production of relatively thin sectioned iron, steel and non-ferrous castings without drying. Moulds for heavy castings produced in

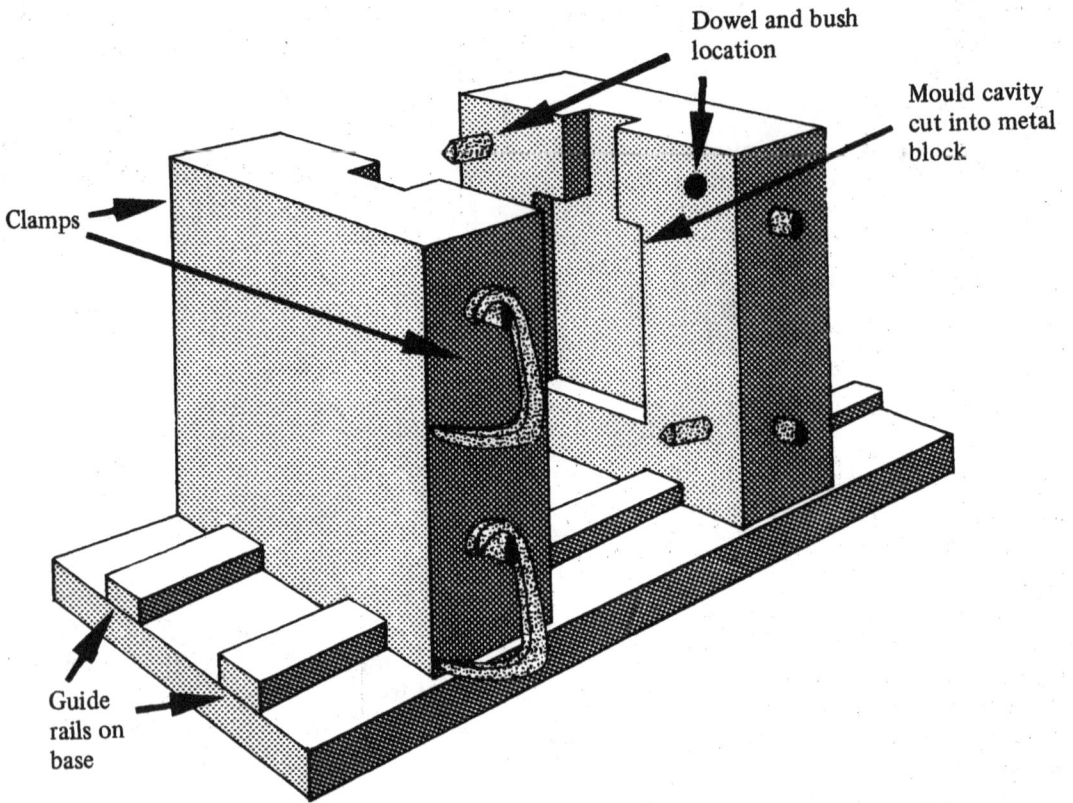

Dowel and bush location

Mould cavity cut into metal block

Clamps

Guide rails on base

Metal Die or Permanent Mould

Chimney

Moist sand loaded into drum

Dry sand
flows over
cone and
through
gap

Gas flame
or other
heat from below

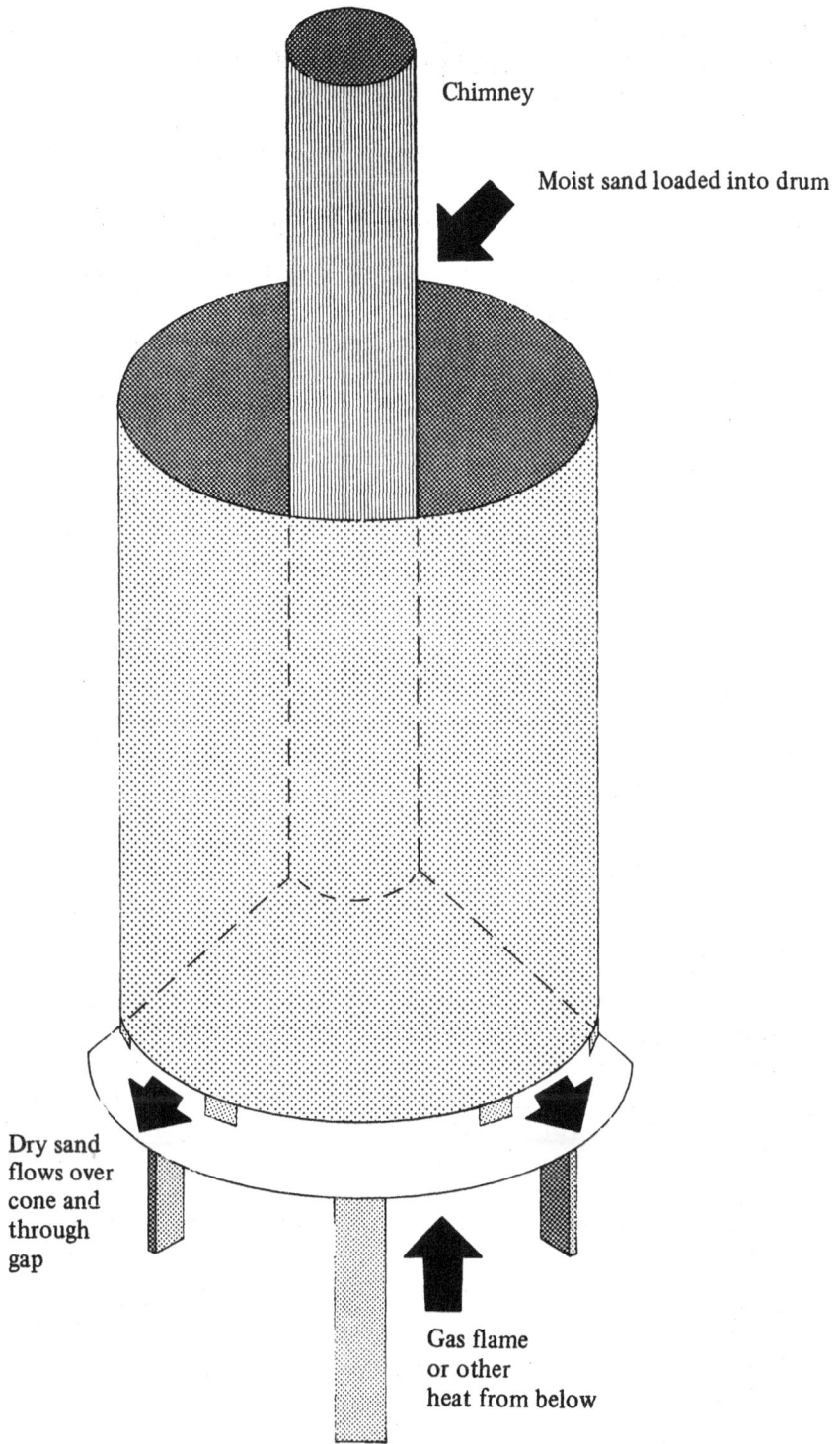

A Simple Sand Drier

15

clay-bonded sand should be dried before use. Drying may be carried out in ovens, or with portable mould dryers fired by gas, oil or other fuel. Dried moulds are frequently coated with clay-graphite paints and re-dried before use.

Clay-bonded sand moulds used without drying are known as green sand moulds.

An intermediate process is suitable for some castings, for which a green sand mould is skin dried by the application of heat from the open mould face. A dry sand layer of a few centimetres thickness is formed. Skin dried moulds should be poured within a short time of drying, since moisture from the undried green sand can soak through to the mould face once more.

Sand Preparation

The mixing of sand for foundry use is most important. It is possible to mix sand, especially sand bonded with natural clay, by extensive work by treading and with shovels and sieves. However, for the most efficient results, sand should be mixed in a roller mill or some other type of high intensity mixer. Such mixing and milling is essential for sand with added synthetic clay.

In addition to the sand and the clay, coal dust is often added for iron castings. Up to about 10% of this material is needed to produce the best casting surface finish. The type of coal should, if possible, have low ash and sulphur contents. Other carbonaceous materials such as powdered pitch and certain oil products may be used instead of coal dust. It is not usual to add coal dust for steel and certain types of non-ferrous castings.

Foundry sand can be used repeatedly. Used sand has to be sieved to remove lumps of cores and pieces of metal. The sand is then cooled and mixed with more water, new sand (about 10% of the weight of metal which has been poured) and more clay and coal dust as necessary. To economise on mixing and on additions, such prepared sand can be used as a facing sand for a layer 2 cm to 5 cm thick in contact with a pattern to form the mould face. Old sand which has been less carefully prepared is used as a backing sand.

In preparing a mixture of sand for moulding the proportions of new sand, additional clay if any is required, coal dust and moisture must be carefully and accurately weighed or measured.

It is possible to obtain laboratory equipment to test and determine sand properties. For simple foundries this equipment may not be justified, but careful control of mixing, of composition and of water content should always be exercised.

Types Of Sand

Most foundries use silica sand; however for high quality steel castings and for some special cores, sand based on heat resisting minerals may be used, despite their higher cost in countries which have to import them. Zircon, Chromite and Olivine sands have particularly valuable foundry properties, and should be considered for use when they are available locally.

Hand Moulding

The compaction of the sand around the pattern is most simply performed by ramming with a wooden hand tool, taking care not to damage the pattern. As an alternative to hand ramming for large moulds much labour can be saved by using pneumatic rammers if compressed air is available.

The sand is compacted uniformly and hard, to give as rigid a mould as possible.

Moulds should be permeable to the gas which is formed on pouring; if the sand is fine grained it is necessary to vent the mould using a wire pricker to make gas escape passages.

If the pattern is not mounted on boards, the hand moulder has to create the mould joint by cutting the sand, ram both halves, and strip the pattern from the mould. He must lay any cores, clean the mould by blowing with compressed air or bellows and close the mould together ready for pouring.

Skilled hand moulders can make castings of great complexity from relatively simple

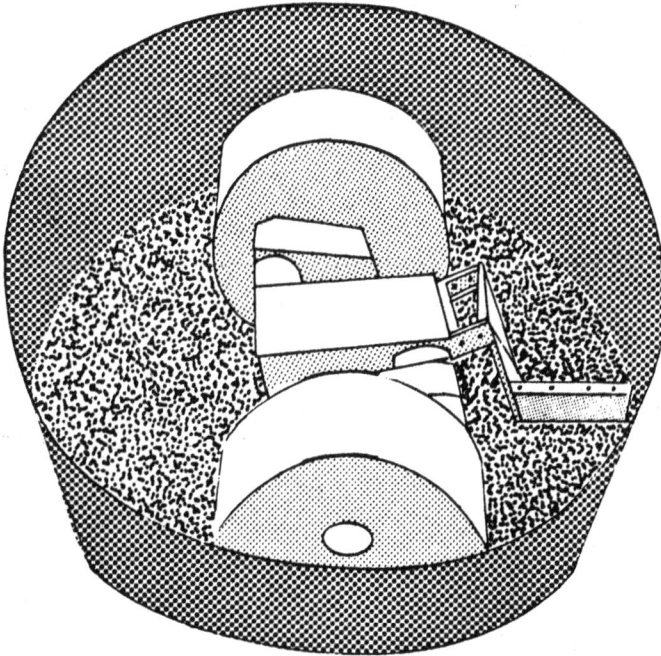

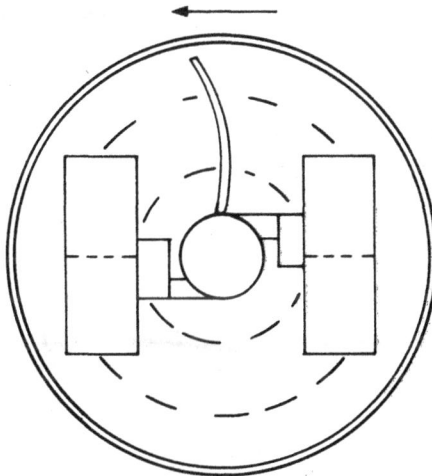

Plan view

The rollers work the sand against the base of the mill,
whilst the plough scrapes the sand from the walls and the base.

Sand Mill

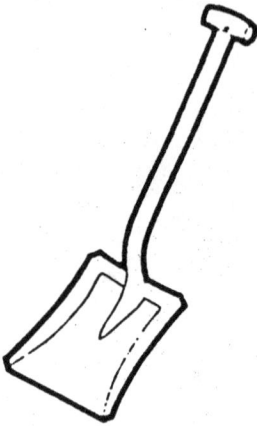

Sand shovel

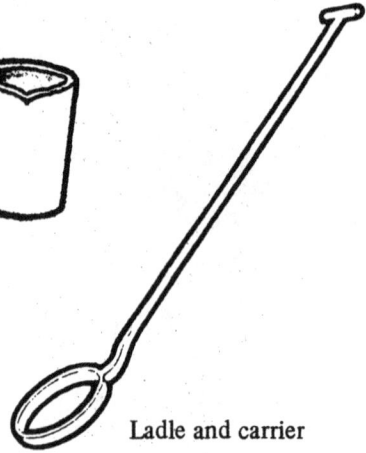

Ladle and carrier

Bellows for cleaning
loose sand from moulds

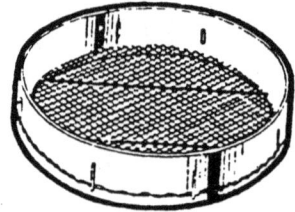

Sieve for facing sand

Trowels for cutting
sand

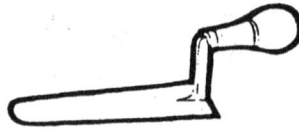

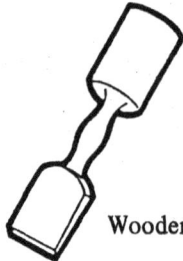

Wooden hand rammer

Some hand moulders tools

patterns, using three or more section moulds to avoid the need for cores, or sometimes using templates or sectional patterns to make uniform mould shapes by hand.

The making of complex castings by hand moulding requires two or more years of training; however simple hand moulded castings can be produced after a few weeks' experience.

The hand moulder's tools include a rammer, shovel, sieve, a fine trowel for cutting the sand, vent wires, lengths of tube for cutting pouring holes etc. Dry powder (for example powdered bone dust) is dusted on patterns and joint surfaces to stop the moulding sand sticking and not separating cleanly.

The moulder must also make the runner channels in the mould when these are not part of the pattern equipment. Loose wooden pieces should be used to mould the ingates into the casting cavity. The runner channels and feeders may be cut with a trowel, or moulded from standard loose pieces kept with the moulder's equipment. A cut sand surface will not be as smooth as a moulded surface, and even if it is smoothed with the trowel may entail the risk of producing sand inclusions in castings.

Machine Moulding
Mechanised casting production needs moulding machines. Moulding machines may jolt sand into position or squeeze the sand round the pattern. Both methods are often combined in one machine.

Most simple squeeze machines are not useful for moulds in which each half is more than about 150–200 cm deep although special high pressure hydraulic machines have been designed for deeper moulds.

Usually moulding machines are installed in pairs, one to make top half moulds and one to make bottom half moulds. There are types of moulding machines which can produce complete moulds on one machine. It is also possible to use one machine to make bottom half moulds, and then later to use the same machine with the other half pattern to make the top half moulds.

Moulding machines usually require a supply of compressed air (e.g. at 6 atmospheres), although some types operate with self-contained hydraulic pumps and require only to be connected to an electrical supply.

Small moulds can be made on simple moulding machines, in which the squeeze pressure is applied by hand with a long lever, using neither pneumatic nor hydraulic power assistance.

Even the simplest moulding machines provide a mechanised means of stripping the mould from the pattern and if necessary of turning the mould over during production.

Although special purpose moulding machines can be very complex pieces of equipment, a well equipped workshop should be able (with some assistance at the design stage) to produce a simple machine from structural steel, iron castings, and hydraulic or pneumatic cylinders and valves.

One special type of moulding machine is the sand slinger. Sand slingers contain impellors with spinning blades to hurl the sand at high speed into the moulds. Sand slingers are expensive machines, used for large castings, and not usually suitable for small scale foundries.

It is possible to purchase automatic moulding machines which produce very high quality moulds at extremely high rates — several hundred complete moulds per hour. Such a moulding machine on its own has no value; it must form part of a complete mechanised foundry system, including sand preparation equipment and mould and sand handling conveyors. The maintenance and the capital costs of such equipment are high and are only justified when there is a very large and assured market for mass production of castings.

Even simple moulding machines are only suitable for repetition production of batches of 20 to 50 or more moulds at a time. A moulding machine should be regarded as a labour saving device which is capable of producing high quality castings with less skill

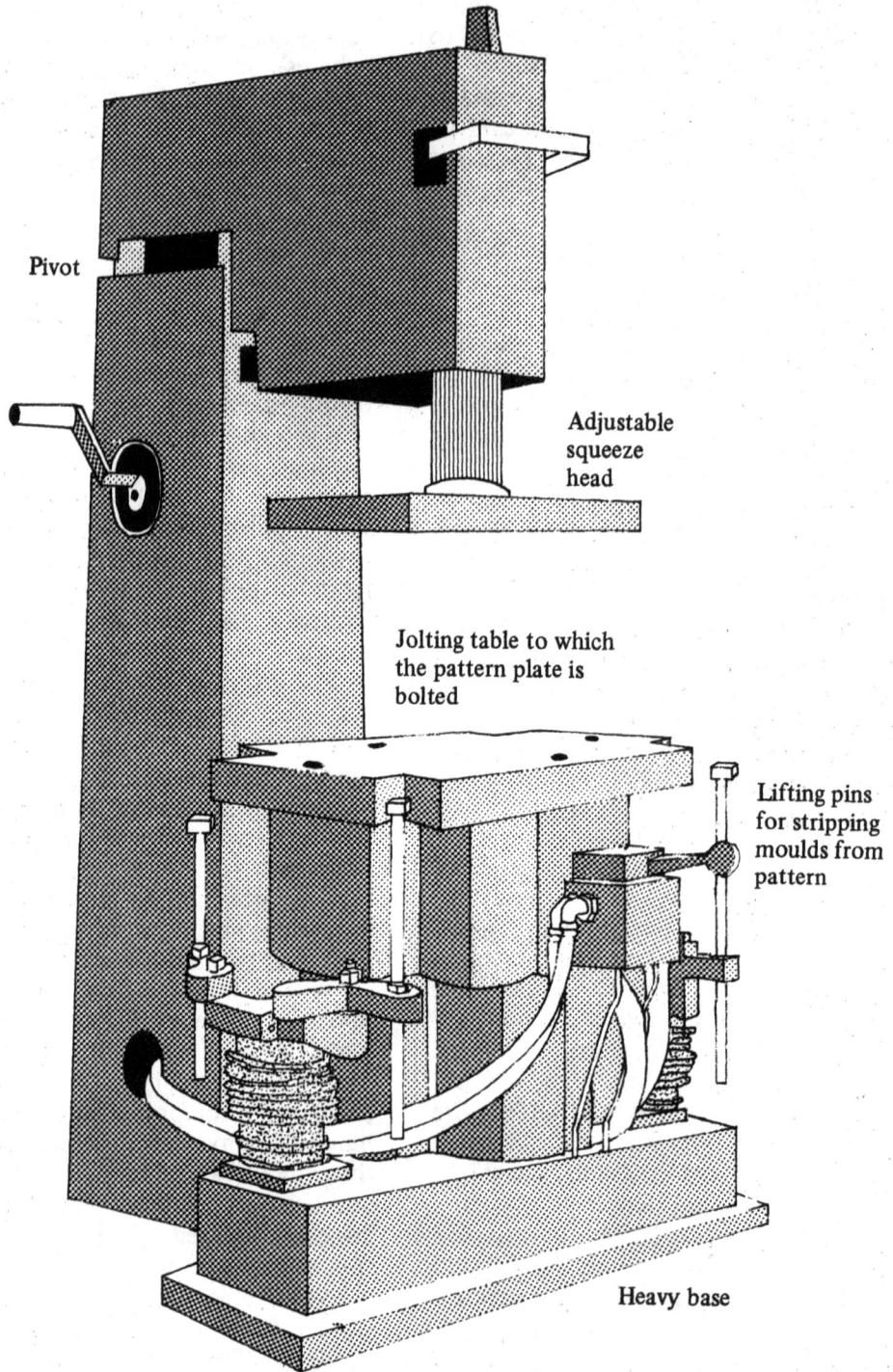

Pivot

Adjustable
squeeze
head

Jolting table to which
the pattern plate is
bolted

Lifting pins
for stripping
moulds from
pattern

Heavy base

Jolt – Squeeze Moulding Machine

than is required by a hand moulder. Nevertheless for any moulding machine the investment is relatively high, and is not usually justified unless there is a sure demand for the castings, and sufficient sand, metal, and space to ensure that the machine is thoroughly utilized.

It is possible to learn to operate moulding machines with a few hours' practice, whilst the production of a mould by hand ramming is a much more skilled operation.

The CO_2 Process

In recent years many techniques have been developed to supplement the use of traditional foundry sands. These techniques involve using sand bonded with materials other than clay.

One such method is the CO_2 or carbon dioxide process. Clean dry sand is thoroughly and mechanically mixed with between 3% and 7% of sodium silicate (water-glass). This material is available for a number of industrial processes. The special grades produced for foundry purposes are the most suitable, but are not always essential. The best ratio of silica to soda is 2 : 1 at a specific gravity of 1.4 to 1.7.

The mould is rammed with the mixed sand and then the water-glass is caused to harden by passing CO_2 (carbon dioxide) gas through the mould. CO_2 gas in cylinders is available for a variety of purposes as well as for foundries.

It is necessary to use a reducing valve from the cylinder in order to control the pressure of the CO_2 gas to about 1.5 to 2.0 atmospheres. The gas is fed into the sand through thin metal tubes or under a board or plastic cover. Care has to be taken not to waste CO_2 gas. The gassing time depends upon the size of the mould or core — from half a minute up to about five minutes. About one or two cubic metres of CO_2 gas are needed for every 100 kg of sand — or 1 kg for every kg of sodium silicate.

Sometimes CO_2 sand mixtures include a small amount of clay or other bonding material so that the mould may be removed from the pattern without collapse before hardening; in other cases the mould is gassed while still in contact with the pattern.

Water-glass and carbon dioxide are relatively cheap raw materials. The sand cannot readily be re-used. Water-glass bonded moulds may also be hardened by special chemicals (esters such as glycerol acetates) or by stove drying. However CO_2 gas if available is generally more satisfactory. The CO_2 process is suitable for all sizes of castings and generally requires less skill than clay-bonded sand moulding. It may be used for cores as well as for moulds.

CO_2 moulds and cores often produce a rough casting finish and for high quality work should be painted with carbonaceous material (graphite mixed with clay or other bonding agents suspended in water or alcohol. The water must be dried or the alcohol burned off before closing the moulds). CO_2 castings often require more work at the casting cleaning stage than green sand castings.

For some castings a CO_2 sand layer of 5 cm to 20 cm may be backed with a clay-bonded green sand. Such moulds are cheaper, although less strong and rigid than full CO_2 moulds. They should not be left to stand for long periods before casting as moisture can soak through to the CO_2 layer and so spoil the hard surface.

Air-set Moulding

A useful method for making a variety of large and of small moulds without the use of skilled labour or expensive machinery is the use of chemically bonded air-setting sand. Clean, dry, clay-free sand (with no shell or limestone contamination) is mixed with about 1½% to 2% of special chemical resin and a hardener, in such proportions that within a few minutes a hard bond is formed. Before the resin hardens the sand flows easily and needs only a minimum of packing or ramming around the pattern. The chemical agents may be mixed with the sand in continuous screw feed mixers; these are usually relatively simple pieces of equip-

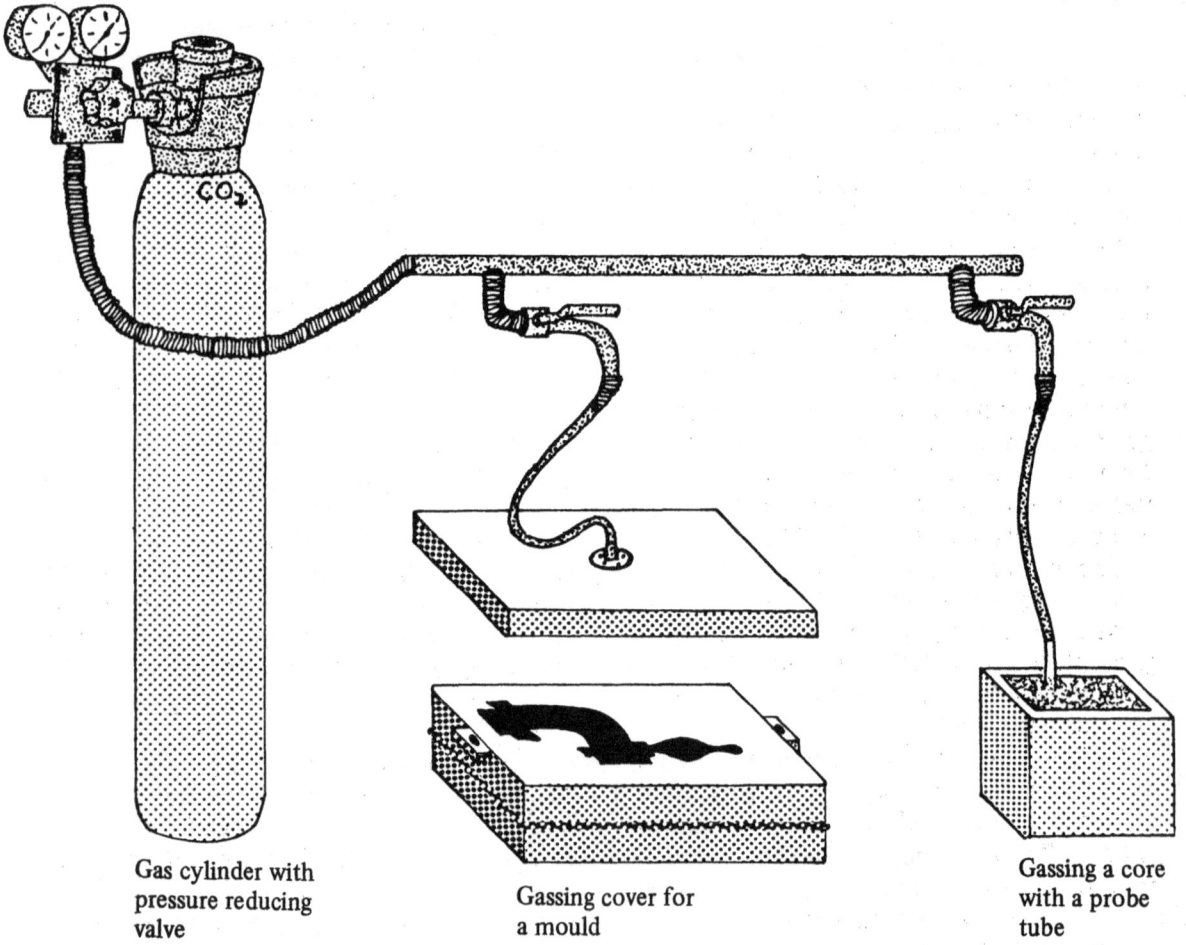

Gas cylinder with
pressure reducing
valve

Gassing cover for
a mould

Gassing a core
with a probe
tube

Carbon Dioxide (CO$_2$) Process Gassing Systems

ment, available from many manufacturers.

An important advantage of the air-set process is that the capital cost for a moulding installation, particularly for large castings, is very much less than the capital cost of sand-milling plant and moulding machines for clay-bonded sands. The skill required to operate the process and the skill required to maintain the equipment is also significantly less. On the other hand it is necessary to purchase the special and expensive chemicals.

It is necessary to choose resins which are suitable for the metals to be cast. Some types of resin deteriorate rapidly if they are stored in very warm or in very cold conditions, so that care must be taken to purchase limited quantities at a time, and of a type suitable for the climate.

An advantage of the process is that simple wooden patterns can be used. Cores can also be produced on the same equipment in the same sand.

If the production of fairly large quantities of moulds is necessary, it is possible to set the mixer over a length of roller conveyor along which patterns and moulds can be pushed continuously for filling.

The economics of the use of chemically-bonded sands depends very largely upon the cost of resins and the ratio of metal poured to sand used. By careful design of the patterns and moulds, avoiding the use of unnecessary quantities of bonded sand, and by minimizing the waste, the economics of the process can be improved.

Chemically-bonded sand is used by foundries producing moulds in moulding boxes and also by others which do not use moulding boxes but assemble the moulds from blocks or assemble separate cores, according to the design and construction of the patterns.

Depending upon the type of resin or binder used, it is possible to crush and reclaim and re-use a proportion of the sand. The economics of reclamation depend upon the cost of new sand in each area.

The binder chemicals normally used are based on furfuryl alcohol-phenol formaldehyde resins specially made for foundry use. Other resin systems, and sodium silicate (water-glass) hardened with special catalysts are also used. Mould curing times can be varied by selection of chemicals, by control of the temperature, and by varying the hardener or catalyst addition. Curing times from as little as 4 minutes up to as much as 1 hour can be obtained.

Other Moulding Processes
The moulding processes considered above comprise those which would normally be considered by most foundries. However mention can be made of some other moulding processes:

Investment Casting
The lost wax or investment process is an ancient and well known method of producing castings. A pattern is produced from wax, often in an accurately made metal die. The wax is coated by dipping (investment) in layers of clay, plaster or special refractory. After the plaster has dried the mould is heated to melt the wax. The wax runs out and molten metal is poured into the resulting cavity. This process is expensive in materials but produces very fine and accurate castings. It is often applied for jewellery and sculpture and when exceptionally accurate small engineering castings for jet engines, sewing machines, and other such products are required.

Shell Moulding
Shell moulding requires the use of iron or steel patterns. These are heated to about 250°C and then covered (either by dumping or blowing using a special machine) with sand which has been previously mixed with 3% to 5% of a heat-curing phenolic resin. After a few minutes a hardened layer of 1 cm or 2 cm builds up on the pattern; the loose sand behind is tipped away for re-use and the hardened shell is removed when cured by further heating. Two shells are clamped or glued together for pouring.

23

Sand
hopper

Screw Mixer

Pivot

Resin

Catalyst

Prepared
sand

Pumps

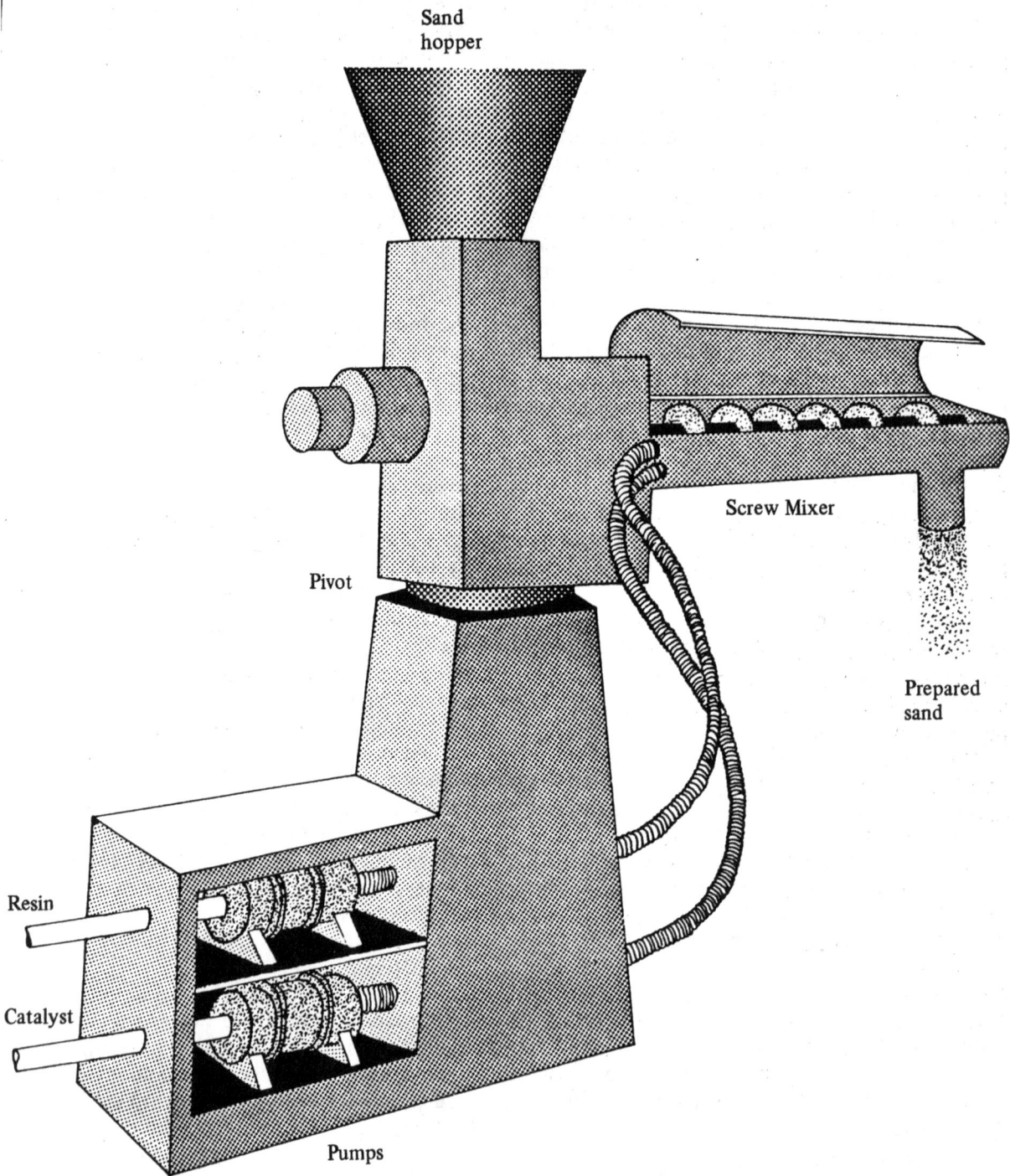

Continuous Mixer-Filler for Airset Process

24

Pattern on Plate

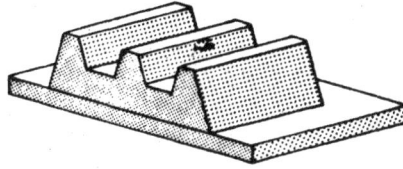

Frame placed over
the heated pattern

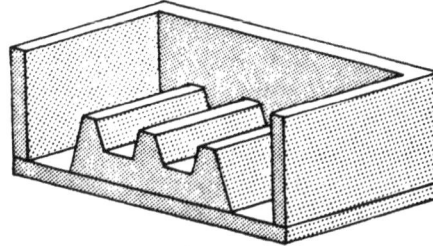

Filled with resin
coated sand which
cures against the
hot pattern

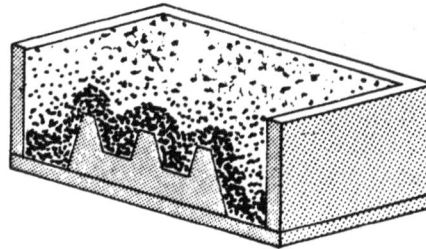

Excess sand removed
leaving cured
shell on pattern

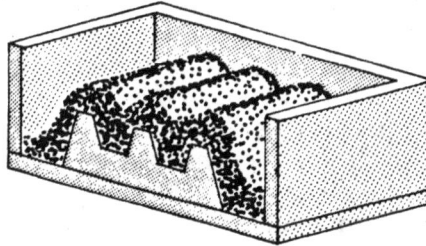

Completed half shell
mould stripped
from pattern

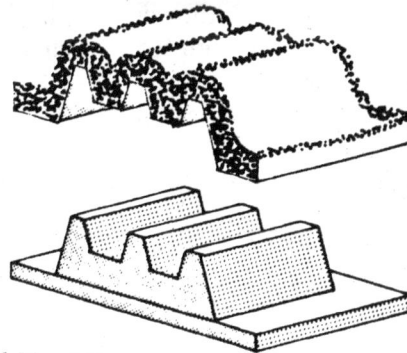

Shell Moulding

The powdered resin may be simply mixed with the sand but it is more effective to use sand whose grains have been coated with resin. It is possible to purchase coated sand: although the coating process is not very complicated, it requires careful technical control and is not usually worthwhile for small quantities.

Shell moulding requires relatively little skill from the operators although it demands high skills from the pattern makers.

The process can produce accurate, smooth, high quality castings in most types of metal. Shell moulds can be stored for long periods without damage and the process can be mechanised in many ways for high production.

Shell moulding is not a cheap method, especially if resin coated sand has to be purchased from a distance. It is suitable only when relatively highly priced castings are to be made which justify the cost of the resins and the expensive metal patterns.

Plaster Moulds

Castings can be made of aluminium and other low melting point alloys in moulds made of carefully dried plaster. These moulds are not re-usable. Castings produced in this way are accurate although the process is expensive. Special types of plaster are available with which it is possible to make iron and even steel castings when exceptional accuracy is required.

Cement Sands

Sand may be bonded with cement. Ordinary Portland cement may be used, although for thick section castings or those that must be poured at a high temperature, heat resisting cement (ciment fondu) is necessary. About 7% to 9% of cement and 6% to 7% of water is mixed with the sand. Before pouring cement moulds must be carefully dried. Cement-bonded sand moulds are strong and are suitable for producing heavy castings; however the sand does not break down easily and this can cause problems with cleaning the castings and with cracking of

thin metal sections. The process is not used by foundries producing small castings, but is one of the processes to be considered for heavy products.

Loam Moulding

A variation of the dry sand moulding process for large castings is the use of loam: this is sand mixed with a very large amount of clay such as fire clay, so that it can be applied by hand in a plastic form against a pattern. Loam moulds are thoroughly dried before pouring. This is the traditional process for making heavy castings (pipes, bells etc.); it demands a high degree of skill and training and is not likely to be useful for small scale foundries in normal circumstances.

Other Binders

With recent developments in chemistry a great variety of compounds has been used for bonding sand for foundry purposes. Many of these chemicals are expensive and present few advantages over conventional materials.

One interesting experiment has been to use ice. Sand is mixed with a little water and the mould is frozen. The metal is poured before the frozen mould has had time to collapse and good castings can be produced. However the necessary high capacity deep-freezing apparatus is very expensive to buy and to operate.

Un-bonded sand may also be held in place by vacuum suction between special thin plastic films. This (V) process is expensive to engineer and is not usually suitable for small scale foundries. It is used for repetition production of large flat castings.

Mould Assembly and Pouring

Moulds must be correctly prepared for pouring, by whatever process they have been produced. They should be inspected for damage and any loose sand or dirt blown away with compressed air or bellows. The cores are laid in the core prints which hold them in the correct position. The top half of the mould is closed carefully and slowly by hand or with a crane (which must be equipped with slow speed controls).

Moulds must be clamped securely together or a heavy weight must be laid on top in order to prevent the molten metal lifting and parting the two halves of the mould. Pouring basins in the tops of the moulds must be kept clean and moulds should not be allowed to get damp before pouring. Pouring of molten metal (clean and free of slag, in a clean pre-heated ladle, at the right temperature) should be steady and fast enough to fill the moulds without entraining of air or scum from the metal surface. Considerable practice is needed in order to pour metal correctly and safely.

CHAPTER 4

COREMAKING

Certain types of casting can be produced without the need for cores. However many designs contain internal cavities or undercuts which cannot be made by normal moulding methods. Very few foundries can operate without some core producing facilities.

The bonding of sand to form cores is similar to that which is required in order to form moulds; many of the same processes are used. There are, however, some differences. A core has to be harder and stronger than a mould since it must be handled and perhaps stored, handled again and laid into the mould by hand. Cores are often almost completely surrounded by molten metal and as the metal solidifies and contracts the core must break down in order not to set up stresses or crack the metal. Cores must not be bonded with chemicals which produce large quantities of gas when subjected to heat since gas bubbles can pass into the liquid metal and form blow holes on solidification.

Core Binders

Cores may be made from clay-bonded sand. However clay-bonded cores are very fragile and require considerable skill in manufacture and support in handling. The use of green sand and dried sand cores has decreased considerably in recent years.

The CO_2 process using water-glass and carbon dioxide gas is an extremely useful coremaking process. The technique is described in Chapter 3. CO_2 process cores can be made with special sand additives to reduce the casting cracking problem which sometimes occurs.

A number of other chemical systems have been developed for producing cores which harden rapidly in their core boxes without heat. A number of chemical resin systems based on phenol formaldehyde, urethane, furfuryl alcohol or other resins harden on the passage of amine gas or of sulphur dioxide. Other resins harden by the action of catalysts mixed with the sand just before coremaking, as with the air-set moulding process. Many of these chemicals are expensive and some present health hazards so that they can only be used with special equipment. Smaller foundries should usually consider only the CO_2 process or the air-set mixer-filler process for cold setting coremaking.

The hot box process needs a heated metal core box: the sand is mixed with resins which cure on the application of heat. This process produces solid cores but in other respects it is similar to the shell moulding process which may also be used for making cores. After 1 cm to 1½ cm of sand has cured shell cores may be emptied out so that hollow cores are produced.

Both hot box and shell cores have to be made on special coremaking machines with heaters for metal core boxes. These processes are convenient and require little operator skill, but the equipment and core-box making and tooling costs are high, so that the processes are usually restricted to foundries requiring high production rates.

Many cores can be made with "natural" binders. Certain vegetable oils, such as linseed oil, cottonwood oil and others, when subjected to heat (temperatures of the order of 250°C), become hard and strong.

Cores are made in simple wooden or metal core boxes but complicated shapes may require support on sand or metal formers until they are baked.

Other natural organic materials are used alone or in conjunction with hardening oils and include starch, flour, molasses, or sugar derivatives. Many types of flour when mixed with water provide a suitable medium for bonding sand to produce cores. A disadvantage is that starch-bonded cores tend to produce large quantities of gas on pouring. Frequently a mixture of starch and oil is used to give a combination between a workable sand before curing and a hard and strong core after heating. It is possible to purchase specially prepared core binders and oils. Some of these oils are based on mineral oils and on fish oils as well as on natural vegetable oils. It is worth carrying out experiments with locally available oils as these may provide more economical core-binders than purchased, specially produced, proprietory materials. A typical mixture might contain 2% oil, 1½% starch, and 2% to 3% of water. Such cores should be cured by baking at 250°C for approximately 45 minutes in an oven.

These oil-bonded sand mixtures may also be used for moulding when particularly strong moulds are needed.

The mixing and preparation of coremaking sand must be carefully carried out. It is usually unsatisfactory to attempt to mix sand by hand, and a small sand mixer of the type used for moulding sand, or even a concrete mixer type, should be used. The composition of core sand mixes must be carefully controlled either by weighing or by measuring the ingredients. Cores which are cured by heat or by the passage of CO_2 gas must be made under controlled conditions. CO_2 process sand hardens slowly in the air, so that to avoid waste it should be mixed in small batches as it is required.

Mixers must be cleaned after use to avoid contaminating other batches.

Coremaking

Cores may be made in metal, wooden or plastic core boxes. These core boxes are part of the pattern equipment for the castings.

The simplest method of making cores is to ram the sand into the core box with a wooden rammer. Many cores may need reinforcement with wire or nails in order to provide internal support. Coremaking for complicated shapes is a skilled process requiring several months of training, although simple cores can be made after a few days or weeks of practice.

An alternative method of producing cores is to blow the sand into the core boxes. Core blowing machines can be bought which are suitable for both hot and cold core making processes. A range of machines is available from simple manually-operated blowers up to fully automatic equipment for the production of intricate cores on a large scale. Such coremaking machines require compressed air, power or gas services and maintenance, and can usually only be justified for repetition castings in conjunction with mechanised mould production.

Cores that have been bonded with oil, starch or some resins must be cured before use. Core stoves may be fired by oil, by gas, by coke, wood or other suitable fuels. It is important that air should be allowed to circulate within the stove since the curing process is by oxidation as well as by the application of heat. It is necessary for the core stove to have reasonably accurate temperature control.

Core Assembly Moulding

Coremaking methods are sometimes used to produce complete moulds. The mixer filler or air-set moulding process described above is one such example, but most coremaking processes can be used for moulding with the advantages of flexibility, rigid moulds, easy stripping, and absence of the need for moulding boxes. Moulds made from assemblies of cores have to be securely clamped and sealed together before pouring the metal.

Coremaking sands are usually more expensive than moulding sands, so core-moulding is only used when there are definite technical advantages.

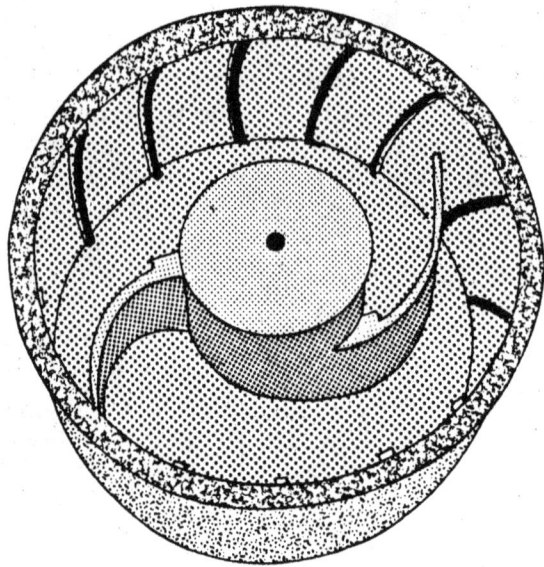

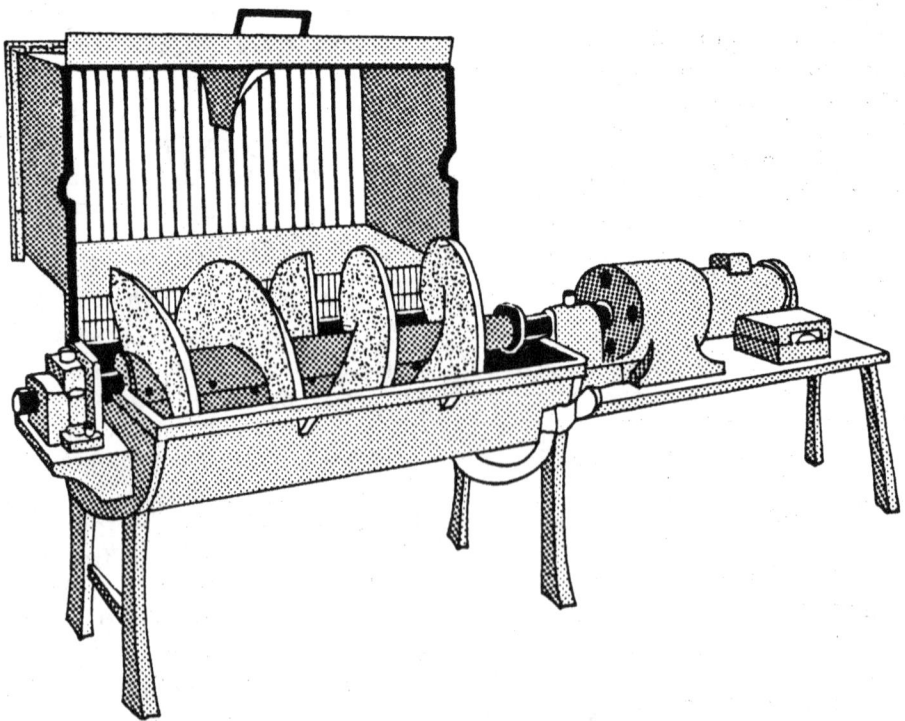

Two types of Core Sand Mixer

SUMMARY OF FEATURES OF MAIN SAND BINDER SYSTEMS

	Clay-bonded sands		CO_2 Process	Air-set Process	Oil & Starch Binders	Shell Process	Hot box Process
	Green sand	Dry sand					
Material costs	Low	Low	Moderate	High	Moderate	High	High
Pattern or corebox costs	Low	Low	Low	Low	Low	High	High
Costs of machines	Low Hand High Machine	Low Hand	Low	Moderate	Low Hand	High	High
Level of skills	High Hand Low Machine	High	Low	Low	Moderate	Low	Low
For small moulds	Yes	Yes	Yes	Yes	Yes*	Yes	No
For large moulds	No	Yes	Yes	Yes	No	No	No
For small cores	No	Yes*	Yes	Yes*	Yes	Yes	Yes
For large cores	No	No	Yes	Yes	Yes	No	No
Re-use of sand	Yes	Yes	Some	Some	Some	No	No
Need for heat to cure	No	Yes	No	No	Yes	Yes	Yes
Accuracy and finish quality	Fair	Fair	Fair	Fair	Low	Good	Good
For high production	Yes	No	Yes	Fair	Fair	Yes	Yes

* Possible, but not generally used

31

CHAPTER 5

MELTING CAST IRON

"Cast iron" is a general term used to cover many different types of material. The most common form of cast iron is grey iron which is used for most engineering applications. Malleable iron is a different material often used for pipe fittings and other castings which need toughness and ductility. This material requires prolonged heat treatment and a high degree of technical control.

SG iron or nodular iron also requires good technical control and laboratory facilities. SG iron is a strong and ductile material used for many high duty applications, sometimes replacing steel. It is made by adding magnesium alloys to molten iron of high purity (made by melting special raw materials, or from steel scrap and refining the molten metal). Special processes are needed to add magnesium safely and efficiently.

Most new iron foundry enterprises will produce grey iron castings only, at least initially.

Technical Requirements

The properties of cast iron and the quality of iron castings depend upon the composition of metal. Grey iron, although a common engineering material, is an extremely complicated alloy containing carbon, silicon, manganese, sulphur, phosphorus, and other elements mixed with the iron. The carbon and the silicon content are particularly important. If the carbon and silicon are low the iron will tend to be hard, brittle and unmachinable. On the other hand if the carbon and silicon contents are too high heavy section castings may be weak and soft.

The fluidity of molten iron is improved by high phosphorus and carbon contents, although phosphorus has a weakening effect upon iron castings. Other impurities such as chromium and other constituents of alloy steel can contaminate cast iron and render it hard and brittle. Lead contamination can also weaken cast iron. It is necessary for the manganese and sulphur contents of cast iron to be correctly balanced in relation to the pouring temperature. If there is not enough manganese, castings will be liable to contain slag inclusions and blow holes.

Cast iron of the following composition is likely to be suitable for foundries making a range of general engineering castings.

Carbon content	3.4%
Silicon content	2.3%
Manganese content	0.6%
Sulphur content	0.07% maximum
Phosphorus content	0.5% maximum

For high strength and thick section engineering castings the carbon, silicon, and phosphorous contents should be lower (e.g. 3.2%, 2.1%, and 0.2% respectively) while for thin section castings for stoves etc. the carbon and phosphorus contents should be higher (e.g. 3.5%, 1.0%). Other compositions are required for other types of cast iron — white iron, malleable iron, SG iron etc.

It should be realised that the composition of cast iron may change during the melting process, especially in rotary furnaces and cupolas where a silicon loss and sometimes a carbon loss has to be compensated for.

When pig iron of known composition is used as the main raw material and when the melting process is well controlled, problems due to metal composition are not likely to arise. However when cast iron is produced from scrap or when castings with different requirements have to be produced it is desirable that the foundry should have some access to basic metallurgical knowledge. Not every foundry can afford a trained and qualified metallurgist, nor a laboratory for analysis, but many foundries can arrange to obtain occasional analyses and advice as necessary from universities or technical institutions.

There is no substitute for practical experience and training. Without the services of someone who has worked in foundries before, a long learning period must be anticipated for any new foundry organisation, even if it is planned to make only the simplest types of casting.

The carbon and silicon contents are the most important elements to control. It is possible to carry out shop floor tests which can give a good indication of the carbon and silicon content without the necessity for an analytical laboratory.

Casting an accurately formed wedge shaped piece of iron and breaking it to examine the fracture can tell an experienced foundryman much about the composition of an iron melt. It is also possible to purchase instruments which register the change in temperature as samples of iron solidify. The changes in cooling rate are related to the carbon and silicon content. This equipment is expensive, although cheaper than a laboratory, but it can give accurate carbon and silicon contents which are invaluable to any foundry attempting to produce iron castings to a close specification.

Raw Materials
The best raw material for the production of cast iron is pig iron. Pig iron is made from iron ore: many different compositions and qualities of pig iron are produced suitable for different types of castings.

Grey iron castings may also be produced by the use of cast iron scrap as a raw material. Most cast iron is produced from a mixture of cast iron scrap and pig iron. Castings made from scrap alone will tend to have low carbon content and be hard and difficult to machine. Cast iron scrap is usually available from broken machinery and automobile engines. It is important to ensure that the cast iron scrap is not contaminated — particularly with alloy steels or other metals.

Scrap must be broken into pieces small enough to suit the melting furnace which is to be used.

The composition of cast iron has to be varied (in particular the carbon silicon and phosphorus contents) in relation to the thickness of the casting and the application. High strength cast iron cannot be produced from cast iron scrap containing phosphorus; phosphorus is often present in scrap castings which have been made for road furniture, gutters, gratings, and heating stoves. On the other hand phosphorus improves the fluidity of molten iron, so that this type of scrap is useful if thin sectioned castings are to be made.

It is also possible in some circumstances to use steel scrap for the production of cast iron. Steel scrap requires the addition of carbon and silicon during the melting process. Steel scrap must be carefully selected to ensure the absence of alloys. Even small amounts of chromium plate, of tool steel or stainless steel, can ruin a melt of cast iron.

Unless special melting furnaces (electric induction furnaces) or mixing devices are available, only small proportions (e.g. 5%) steel scrap should be used in normal grey iron melting.

Any necessary silicon additions are made as ferro-silicon. Ferro-silicon is produced for the steel industry. Large quantities of power are needed to produce it, so that it is made in countries with hydro-electric resources.

Synthetic graphite or coke is sold for re-carburising steel scrap in induction furnaces. However re-carburisation can also be achieved using charcoal. Re-carburisers must not contain undesirable amounts of sulphur, nitrogen or moisture.

33

Cupola Furnaces

A large number of different types of furnace may be used for melting cast iron. Of these the coke-fired cupola furnace is the most common. This furnace consists of a shaft, lined with firebrick or fire clay, often with a sand bottom rammed on to doors which open to discharge the residue after a melt.

A coke fire is made in the cupola, and alternate layers of metal (pig iron or scrap) and more coke are charged on top of the bed of incandescent burning coke. Air is blown in through apertures (tuyeres) in the side of the furnace. Molten iron accumulates and is tapped-out through a hole at the bottom. The tapping hole is usually blocked with a sand plug, except when a ladle of iron is being tapped out.

The coke ash and any other dirt is formed into a liquid slag by limestone, which is charged with the coke. The slag is tapped from the back of the cupola periodically during a melt.

A cupola furnace is relatively inexpensive to construct. However the performance of a cupola depends upon its design, and expert advice should be sought if a home-made cupola is being planned. It is essential to provide a fan of sufficient capacity to maintain combustion at a high level and ensure sufficient metal temperature. Instruments to measure the air volume or pressure are very useful.

As an approximate guide a cupola should have an internal diameter (the diameter of the steel shell less a lining thickness of at least 15 cm per side) and be equipped with an air blower as specified below:

Tons Per Hour	Internal Diameter	Volume of Air at 0.1 kg/cm² Pressure
1	45 cm	23 cu m / min
2	60 cm	40 cu m / min
4	75 cm	63 cu m / min
6	90 cm	90 cu m / min

The four to six tuyeres should be set from about 25 cm to 90 cm above the sand base. The stack above the tuyeres should be at least 3 metres high plus a taller chimney above the charging level to remove the fumes from the foundry.

It is also necessary to provide a means of hoisting the charges to the furnace sill.

If the air supply to a cupola is accurately measured and divided equally between two rows of tuyeres, better and more economical performance results. The cost of the equipment to achieve this is likely to be justified by a coke saving in cupolas which melt for more than 4 to 5 hours at a time, at more than 2 or 3 tonnes per hour.

A full size cupola furnace is not likely to be economical for melts of less than 3 or 4 tons. Some foundries using cupolas produce moulds every day which are left on the floor for several days until sufficient have accumulated to justify a melt.

Cupolas melt continuously, and do not produce large batches of molten iron at one time. Depending upon the size of the furnace, cupolas can be made to melt at from as little as 1 tonne per hour up to 30 or more tonnes per hour.

Cupola melting is relatively inexpensive, although it is necessary to obtain supplies of suitable coke. Large cupolas require coke produced especially for the purpose; cupolas of less than about 2 tonnes per hour can use other types of coke such as that used for blast furnaces or even for domestic coke. The sulphur content of the coke should be low. If high sulphur coke is used it will be necessary to add extra ferro manganese to the charge.

Hard charcoal has also been used as a cupola fuel, alone or with coke, although the results are not very satisfactory even on small cupolas.

Cupola coke consumption is about 12% to 15% of the weight of the metal, plus the weight of the initial coke bed for each melt.

The operation of the cupola, should not be undertaken without experience and training. Cupola melting which is not under proper metallurgical control may produce pouring temperatures so low that defective castings are produced. Iron may even

Water spray
dust collector

Charges in

Alternate charges
of metal and coke

Air blast

Incandescent
coke bed
Tuyeres

Slag hole

Metal tap
hole and
spout

Sand bed

Ladle

Drop bottom doors

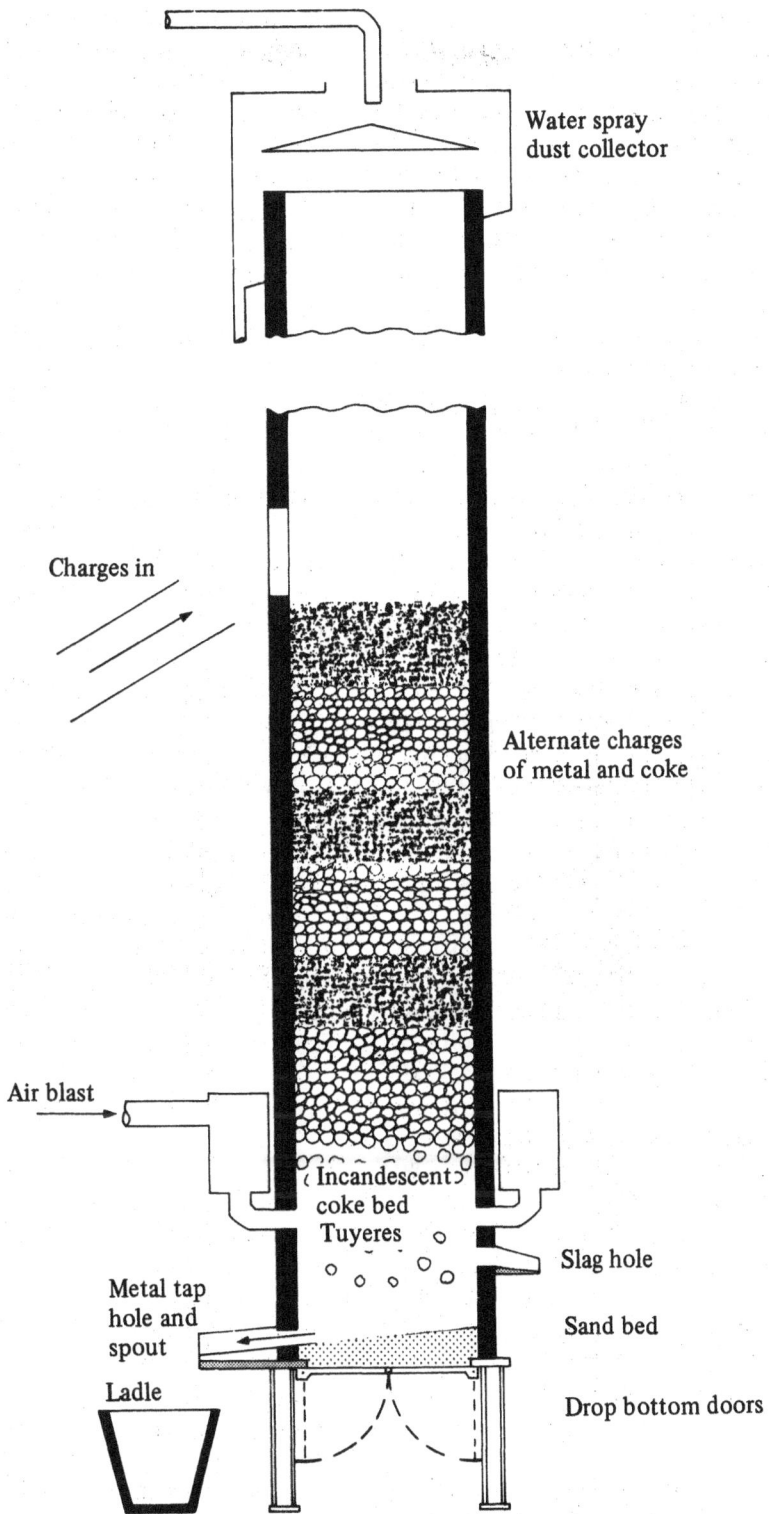

Diagram of Cupola Furnace

35

solidify inside the cupola.

Some small "home made" cupolas give good results, but properly designed furnaces are usually more satisfactory. Very small cupolas ("cupolettes") are often made to tilt on trunnions to help in tapping and slagging. Larger furnaces do not tilt. They require experienced operators to conduct the melt so as to control the air, metal and slag flow.

Gas And Oil Furnaces

In countries where gas and oil are available relatively economically these may be used for melting. Gas or oil-fired rotary furnaces are amongst the most efficient ways of using such fuels. Unlike cupolas which produce molten metal continuously, rotary furnaces are batch furnaces. Many rotary furnaces have a limited size of charging aperture so that large pieces of scrap cannot be used.

Rotary furnaces are equipped with heat exchangers to pre-heat the combustion air with the escaping flue gas, thus saving fuel. These furnaces can be made with capacities from a half tonne up to twenty-five tonnes or more. In rotary furnaces there is a carbon and silicon loss so that pig iron charges have to be used with relatively little scrap.

Rotary furnaces require large quantities of fuel to pre-heat the linings and are therefore most efficient if used for several melts consecutively. Fuel consumption is about 130 to 230 litres of fuel oil per tonne melted.

Some manufacturers of rotary furnaces produce burners which can be used to adapt the furnaces to run on powdered coal instead of oil or gas.

Gas or oil fired crucible furnaces (as described below for non-ferrous metal melting) can also be used for melting cast iron in batches of up to ½ to ¾ of a tonne. Fuel consumption would be approximately 0.3 to 0.6 litres of oil per kilogramme of iron.

Electric Melting

Electric furnaces may be used for melting steel scrap, cast iron scrap or pig iron.

Electric furnaces are expensive in themselves and also require capital expenditure for electrical supply, transformers, capacitors, and contactors. Nevertheless their flexibility and adaptability make them the selected unit for many foundries, even when electrical supplies are not available and it is necessary to install generators. The economics of using an electric furnace depend upon the cost of energy, which can sometimes be offset by the use of cheaper or more readily available raw materials such as steel scrap.

Electric furnace maintenance requires skilled electricians and engineers and expensive spare parts. The lining materials used for electric furnaces can represent a major cost. It is unlikely that the special purpose lining materials required for induction furnaces can be obtained from other than specialist sources.

As an approximate guide to power consumption, a continuously utilized efficient electric furnace will consume between 500 and 600 kilowatt hours per tonne of iron melted. However with intermittent operation power consumption over twice this level is not uncommon.

The power supply must be secure and continuous without risk of interruption. Electric furnace specifications and power supplies should be discussed with electrical supply authorities.

Types of Electric Furnace

Coreless Induction Furnaces

Induction furnaces operating at medium frequency (200 to 1000 cycles) are flexible and convenient and with modern electrical control systems are becoming relatively less expensive. Lower frequency (50 to 60 cycles) induction furnaces should operate continuously and are best suited when only one type of iron is to be melted. These "mains frequency" furnaces are very useful for recarburising steel scrap, and for melting thin sheet scrap and swarf.

Charge through open top

Air blast through
trunnion supports

Spout

Stand

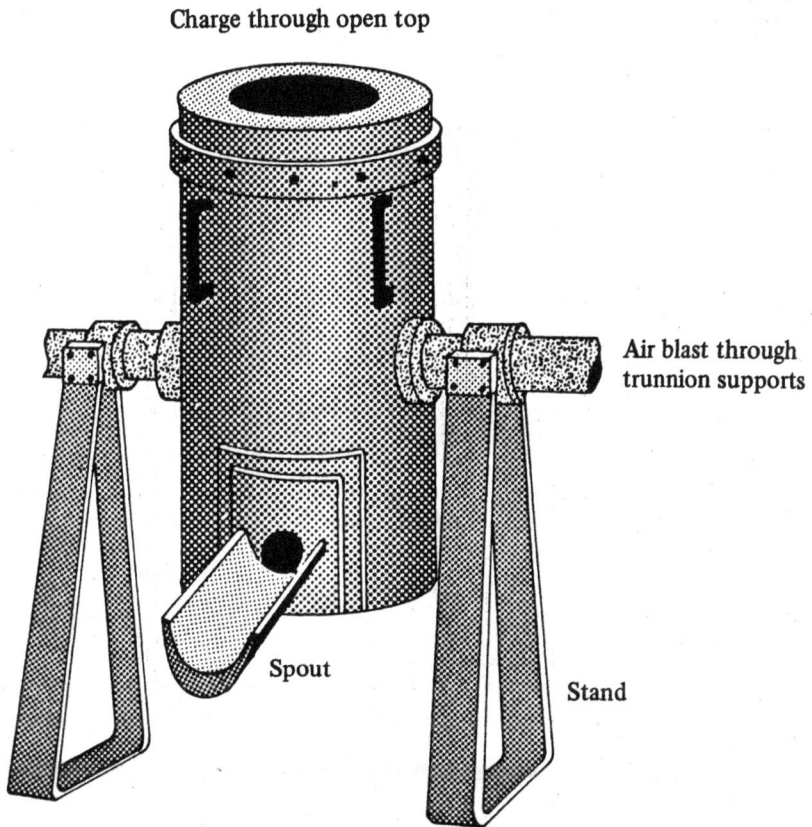

Small Cupola Furnace ("Cupolette")

Chimney

Heated air

Heat
exchanger

Burner

Oil
supply

Roller supports

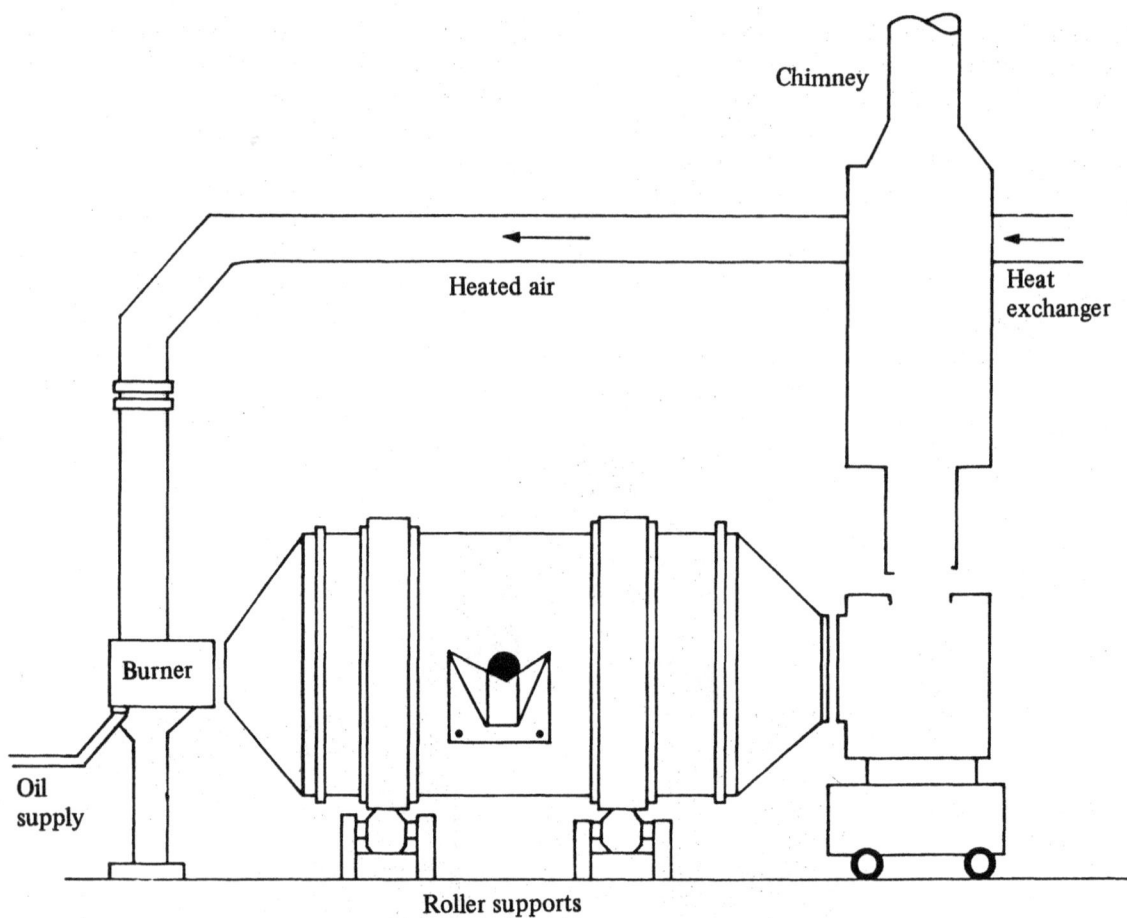

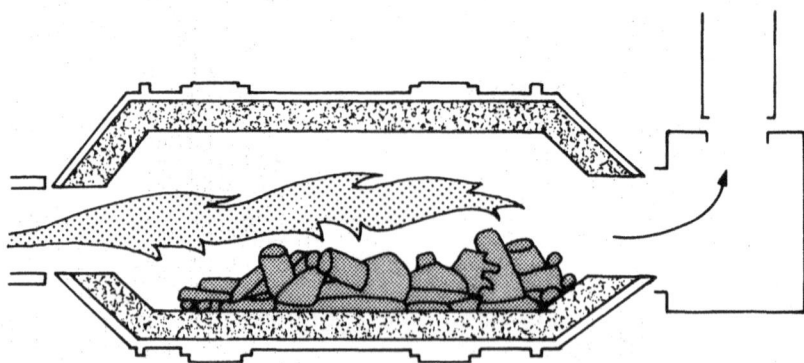

Oil Fired Rotary Furnace

Channel Type Induction Furnaces

These furnaces are cheaper than coreless furnaces but are usually less suitable for melting than for holding molten metal melted until required. They must operate continuously for months at a time without interruption.

Arc Furnaces

These furnaces are suitable for melting many different types of scrap. They are most efficient when used at full capacity. The carbon electrodes used in arc furnaces are consumable and must be purchased, about 5 to 10 kilogrammes per ton of metal melted are required. Arc furnaces are particularly useful for large scale steel melting.

Electric Resistance Furnaces

These are not generally suitable for cast iron although for lower melting point metals they are the simplest type of electric furnace.

Whichever type of electric furnace is selected is bound to be a major investment. This expenditure requires the best technical advice and can only be justified by full utilisation of the furnace. Electric furnaces are not suitable for foundries with no access to technical, electrical and maintenance services.

Refractory Linings

Whenever metal is to be melted and handled it is necessary to have the correct heat-resistant lining materials for the furnaces and the ladles. Refractory lining materials used depend upon the type of furnace.

Crucible furnaces require no further lining. Cupola furnaces and oil or gas fired furnaces must be lined with refractory bricks and compounds. Fire bricks and ramming materials to a wide variety of specification are made for different types of duty. Silica-alumina fire bricks are normally used for cupolas and for rotary furnaces.

In addition to fire bricks it is necessary to have heat resistant fire clay available for the patching and repairing of furnaces. Clay by itself tends to crack and to crumble when dried, and for this reason it is usually mixed with sand or with crushed fire bricks in order to make refractory patching materials. Fire clay mixtures of this type are often used for ladle linings as well as for furnace repairs.

Ladle linings must be most carefully dried and pre-heated in order to ensure that no moisture remains in the lining. If molten iron is poured onto any damp material there is a danger of boiling, splashing or explosion.

Another method of lining ladles is to use sand. Some naturally bonded moulding sands are suitable for lining small ladles. Silica sand bonded with water-glass, either dried or gassed with CO_2 can also give satisfactory results. For larger ladles holding more than 200 to 300 kilogrammes of iron it is advisable to use stronger ladle lining materials such as fire brick or fire clay mixtures.

Electric arc furnace hearths are lined with similar materials to those used for ladles and for rotary furnaces. However, high strength bricks are necessary for arc furnace roofs.

Electric coreless induction furnaces are usually lined with special high purity silica material. This silica is carefully graded and bonded with small proportions of boric acid. Induction furnace linings are necessarily thin and have to be most carefully installed and maintained.

Capacity of Melting Furnaces

The capacity of foundry melting equipment must be calculated to suit the planned foundry output, bearing in mind the following points:

1. Yield

Allowance has to be made for melting the metal which fills the runner system (which in typical iron foundries may be from 30% to 60% of the poured weight) and for some reject castings, as well as metal which is never poured into moulds. Iron foundries making heavy castings can expect a yield of good castings of between 60% to 70% of the metal melted, whilst foundries making light castings may have yields below 50%.

2. Timing

The melting furnaces in non-mechanised foundries are unlikely to operate continuously. Depending upon the method of organisation, the melting furnaces may only work for half of each day or even less than daily. The furnace capacity must be enough to melt all that is needed in the working time.

3. Pouring Weight

The maximum weight of metal to be poured into any mould should also be considered. The metal must be melted in a reasonably short time. Foundries making small numbers of large castings are likely to have to install furnaces with capacity greater than that which would be dictated by their average requirements.

Such considerations should also influence the choice between batch furnaces and continuous furnaces such as cupolas.

CHAPTER 6

STEEL CASTINGS

The production of steel castings is more difficult than the production of iron castings, due principally to the fact that the melting point of steel is higher than that of cast iron so that higher melting and pouring temperatures are necessary.

It is not possible to melt steel in cupolas, nor in rotary furnaces. Steel may be melted in some types of crucible furnace; however most steel foundries use electric furnaces. Before electric furnaces were available steel foundries used cupolas to melt cast iron and then converted the molten cast iron to steel in converters. Such steel-making processes are now rarely used by steel foundries and are generally only used on a larger scale for bulk steel production.

Medium frequency induction furnaces, although expensive, are the most suitable for steel castings. Arc furnaces may be used for large casting production.

If special types of alloy steel are being made — for example for wear or corrosion resistance — it may be necessary to use special basic (Chromite or Magnesite) furnace linings.

Many steels can be made from scrap, although it is likely to be necessary to check the composition in a laboratory and to add ferro alloys or alloying metals if close specifications have to be met. Control of the melting operation and metal composition is even more important for steel than for cast iron.

Steel castings have different shrinkage behaviour from that of iron castings. Pattern design, runner systems and other techniques are different. In particular steel castings require larger feeders to ensure the absence of shrinkage cavities. The yield of good castings to metal melted is therefore usually lower for steel than for cast iron.

One advantage of steel castings over cast iron castings is their weldability, so that they may be repaired if necessary and joined to other components by welding.

Steel castings may be made in green sand moulds (without coal dust), shell moulds, dry sand, CO_2 sand and other types of moulds. Some types of chemical binders may have to be specially selected for steel castings to avoid the risk of surface defects.

Most steel castings have to be heat treated (normalised or annealed) after casting or after weld repair. Heat treatment furnaces fired by oil, gas, electricity or solid fuel, should be well insulated to avoid heat wastage. A typical heat treatment is to heat the castings to 850°C and hold them at that temperature for two hours. The castings are then cooled slowly in the furnace to anneal them, or withdrawn from the furnace to cool in the air if they are to be normalised. Furnaces capable of providing controlled heat treatment at this temperature have to be carefully designed and constructed.

In general the production of steel castings requires more specialised equipment and more complicated technical processes than does the production of grey iron, aluminium, or bronze castings.

An inexperienced small-scale foundry should not attempt to produce steel castings without comprehensive external advice and assistance.

CHAPTER 7

NON-FERROUS METAL CASTINGS

Raw Materials

The easiest way in which to make non-ferrous castings is to purchase ingots of metal of the correct composition for the alloys required, from a refiner or metal dealer. However it is often cheaper to use scrap. The dangers of using scrap metal lie in the fact that there are many different alloys used for different purposes which cannot easily be distinguished by their appearance.

Aluminium

For example aluminium castings should ideally be made from different alloys from those which are generally used for rolled aluminium products. To use scrap which is not scrap casting alloy may produce defects in all but very simple castings. Most aluminium castings are made from alloys with about 5%, 7%, or 13% silicon.

The selection of raw materials has to depend upon the type of casting being made and its requirements. There is the need for access to metallurgical advice to deal with specific problems if high quality castings are to be made using scrap.

Brass and Bronze

The most commonly produced copper casting alloys are brass and bronze. There are many varieties of these but the most usual are the following:

1. Brass, consisting of 70% copper and 30% zinc, or 60% copper and 40% zinc, is used for valves, pipe fittings, decorative products etc.
2. Phosphor bronze, containing 88% copper, 12% tin and ¼% phosphorus is mainly used for bearings.
3. Gun metal with 88% copper, 8% tin and 4% zinc is used for bearings, gears, bushes etc.

Foundries making different types of metal must always take care to store the returns, ingots, and scrap separately to avoid contamination of one grade with another.

Non-ferrous metal casting yields to metal melted are usually lower than for cast iron, figures of 30% to 50% being normal.

Melting Furnaces

Non-ferrous metals may be melted in oil – fired furnaces, gas-fired furnaces, electric furnaces or crucible furnaces.

Crucible Furnaces

For small scale production the use of crucible furnaces is the most common melting method. Crucibles are pots made of fire clay mixed with coke dust or graphite, very densely compacted, carefully dried, and fired at a high temperature. Crucible manufacture is a specialised and skilled business and few foundries produce their own crucibles.

Crucibles are placed in furnaces. The metal to be melted is charged into the crucible, which may then be covered with a lid. The metal melts as the crucible is heated.

When the metal has reached the correct pouring temperature the crucible may be removed from the furnace with special tongs, and then transferred to a carrier from which the crucible is used as a ladle to pour the moulds directly. Larger sized crucibles are made with a lip, and the whole furnace can tilt so as to pour the metal from the crucible into separate pouring ladles.

Separate crucibles should always be used

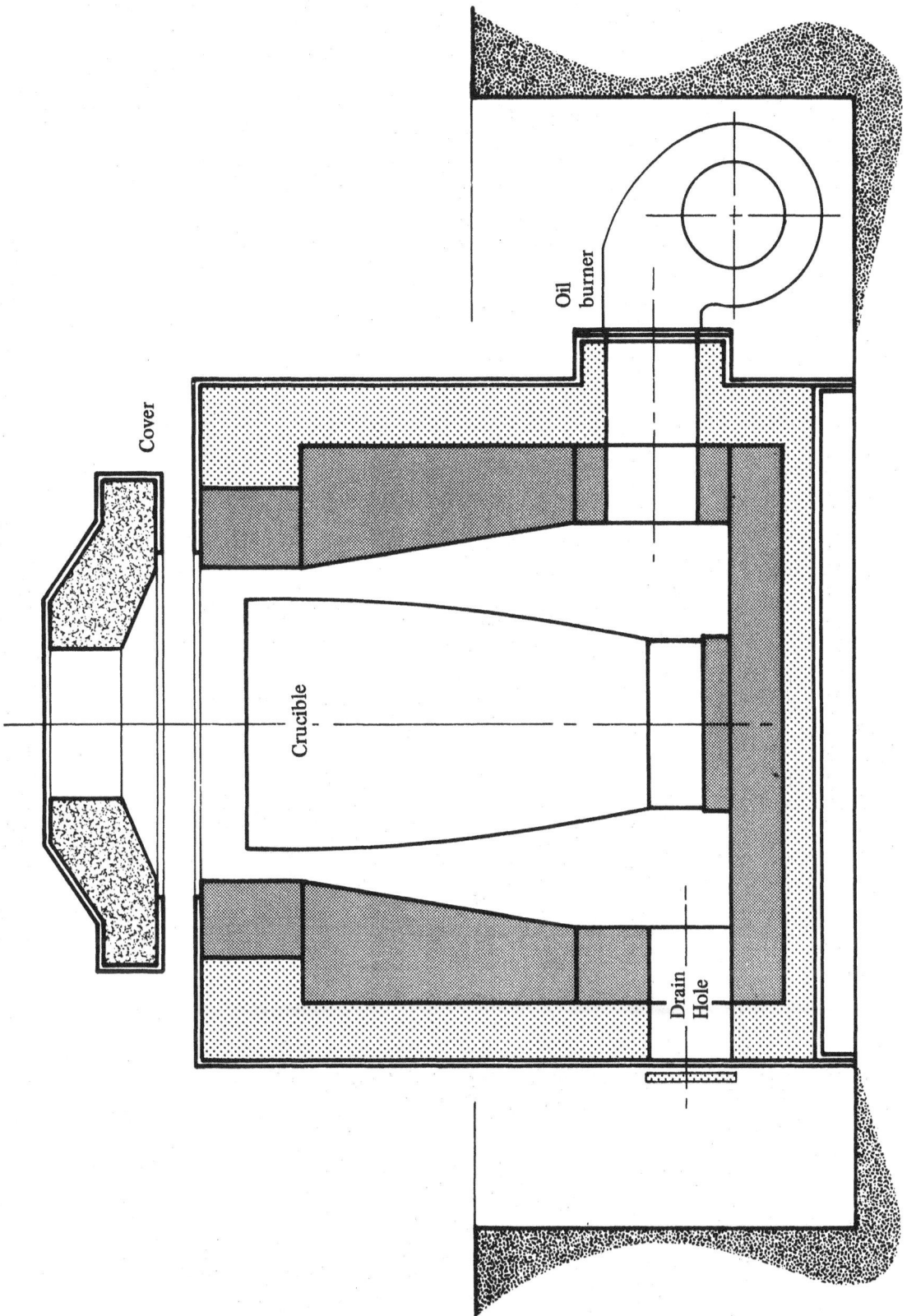

Oil-Fired Crucible Furnace

Tongs for
lifting
crucible

Chimney

Cover

Crucible

extension
to aid
charging

Crucible

Fire

Stand

Fire bars

Draft

Ash

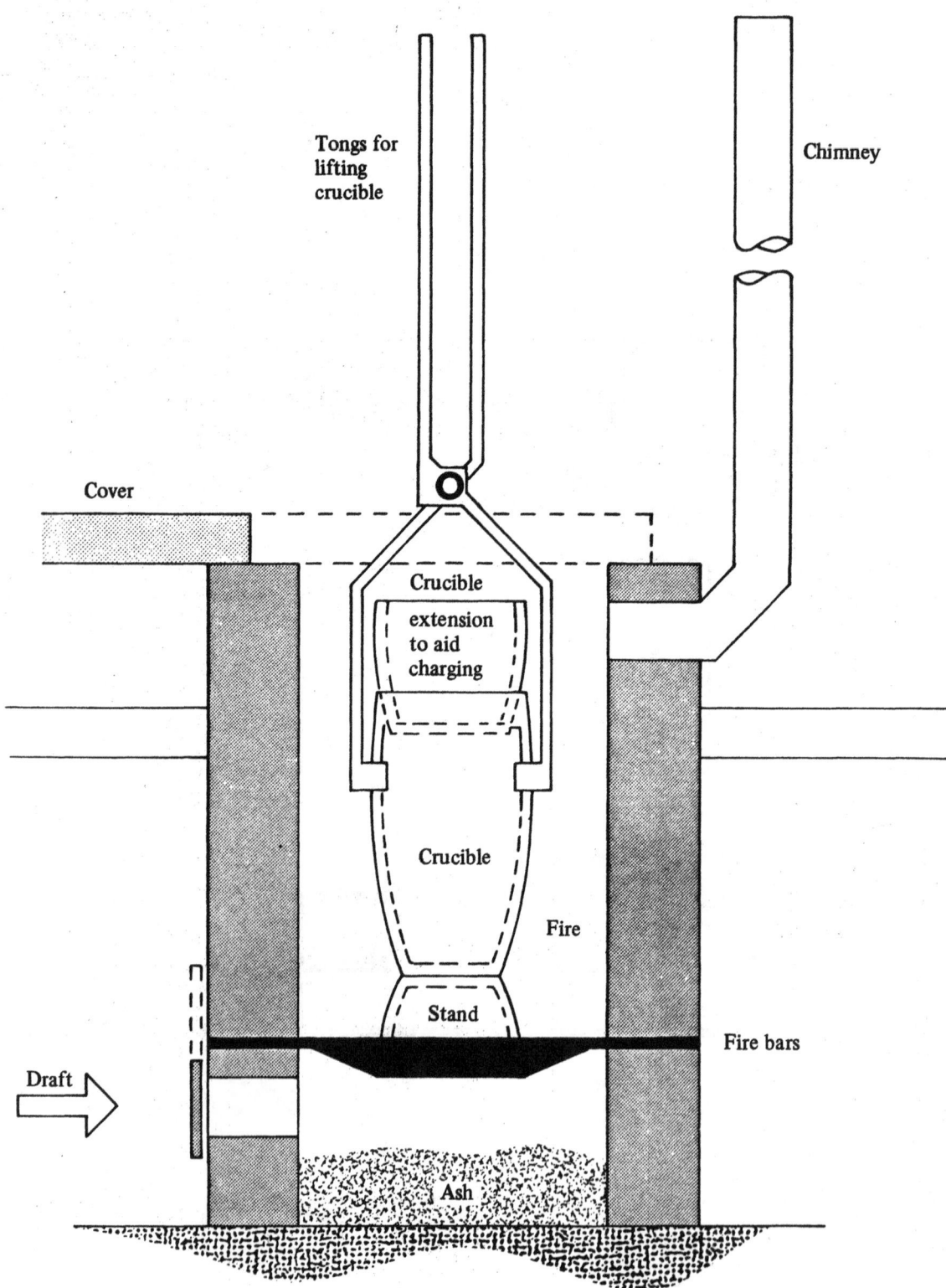

Coke or charcoal fired Crucible Furnace

44

Crucible in Carrier for use as a Ladle

for each type of alloy being melted, to avoid contamination.

Fuel consumption with crucible furnaces depends upon the size of the crucible, the metal being melted and whether or not the crucible is being used continuously. The first melt needs more fuel as the crucible and furnace have to be pre-heated. Typical fuel consumption figures would be 0.2 to 0.5 litres of oil per kilogramme of metal melted. With proper care good crucibles should last for 50 to 100 melts.

Crucibles can also be used to melt metal in coke or charcoal solid fuel furnaces. The crucible stands on bars in the fire, which should be blown with a fan to increase the temperature. (Firing without blowing can sometimes produce a sufficiently high temperature to melt cast iron and can melt aluminium or bronze in properly designed furnaces quite satisfactorily).

Pot Furnaces

For low melting point metals such as lead and zinc-based alloys it is possible to melt in cast iron pots. Such cast iron pot furnaces may be emptied by tilting or by ladling from the furnace. There is no need for furnace lining materials. This type of furnace is commonly used in conjunction with the die casting process using metal moulds.

Electric Furnaces

Electric induction melting is also very suitable for non-ferrous metals. Induction melting units are available in which the metal is placed in crucibles which are themselves placed within an electric induction coil. On completion of melting, the crucible is lifted from the coil and handled as if it had been heated in a conventional crucible furnace. This type of furnace, although relatively expensive, is an extremely satisfactory way of melting small to medium quantities of a variety of non-ferrous metals.

Melting Aluminium

If aluminium is melted in a cast iron pot, it is necessary to coat the side of the pot with clay to avoid metal contamination.

However aluminium is melted it is subject to the risk of scum and dross formation, and to the pick up of gas, both of which cause casting defects. It is possible to add special fluxes just before pouring to reduce this risk. These fluxes, such as aluminium chloride, or proprietary mixtures, help the dross to coagulate into a slag which can be skimmed off, and also liberate a gas which bubbles through the metal and cleanses it from the types of gas which cause the defects.

Melting Bronze

If dirty scrap is used, it may be necessary to flux bronzes, although normally it is sufficient to add a few pieces of charcoal to the melt.

Neither bronze nor aluminium should be left in the molten state for long periods, but should be poured as quickly and cleanly as possible after melting.

Pouring

The control of pouring temperature is very important for the production of good castings in bronze and aluminium. Bronzes are poured at between 950°C and 1000°C, while aluminium is poured at about 700°C.

If high quality castings are to be produced, the cost of a simple pyrometer to check metal temperature is likely to be repaid in a short period by the savings in the numbers of defective castings.

As with all metals, it is very important to avoid turbulence and splashing, especially with aluminium alloys.

Moulding and Coremaking for Non-ferrous Metal Castings

The methods used for non-ferrous castings do not differ in principle from the those used for other metals, although permanent metal moulds (dies) are more suitable for these metals than for iron or steel.

Green sand moulds are normally used for thin-sectioned castings. Care must be taken to ensure that there is not too much water in the sand, and coal dust is not usually added for non-ferrous castings. Since thick-sectioned castings are prone to the formation of surface blow holes due to a reaction between the moisture in the sand and the metal, larger non-ferrous castings are made in dry sand moulds, or CO_2 moulds.

Since there is less heat than in iron castings, problems are sometimes encountered with hard cores, such as CO_2 cores, which do not break down after pouring. Weak core mixes, hollow cores, or green sand cores are often used.

Most non-ferrous metals shrink more than grey iron during solidification. Therefore, it is necessary to use more feeders, and larger feeders, to avoid porosity. The use of chills (see Glossary) is also common in many non-ferrous foundries.

SUMMARY OF CHARACTERISTICS OF MAIN TYPES OF MELTING PROCESS

	For Cast Iron	For Steel	For non-ferrous	Furnace Costs	Lining Costs	Skill needed	For very low output	For high output
Cupola	Yes	No	No	Mod.	Mod.	High	No	Yes
Rotary	Yes	No	Yes	Mod.	Mod.	Mod.	Yes	Yes
Crucible, natural draught	No	No	Yes	Low	*	Mod.	Yes	No
Crucible, forced draught	Yes	No†	Yes	Low	*	Low	Yes	No
Electric Induction	Yes	Yes	Yes	High	High	High	Yes†	Yes
Electric Arc	Yes	Yes	No	High	Mod.	High	No	Yes
Electric Resistance	Yes	No	Yes	Mod.	Mod.	Low	Yes	No†

* No lining costs as such, but replacement crucibles required.
† Possible, but not generally used

CHAPTER 8

CLEANING CASTINGS

After cooling and solidfying in the mould — for from 20 minutes to 24 hours depending on size — the casting is broken out from the sand. This may be done by hand, with hammers, tongs and shovel, or may be mechanised on a vibrating steel grid which shakes the sand out of the mould onto a conveyor for collection and re-use, leaving the castings and moulding boxes free of sand on top of the grid.

The first step after knocking away the loose sand is usually to remove the runners and feeders. Cast iron runners can usually be broken off with a hammer, taking care not to crack the casting. The runners should be designed to break off cleanly with a notch near the ingate into the casting. Steel and non-ferrous castings may need to be cut away with abrasive cutting discs, files, hand or mechanical saws or — in the case of large steel castings — oxy-acetylene cutters.

The next operation is to remove any adherent sand. For small scale production this can be achieved by hand with a wire brush.

When there is a large quantity of castings to be cleaned, and when it is necessary to remove all traces of sand, a shot-blasting machine may be used. There are two types of shot-blasting machine. In one type the shot or abrasive is blown pneumatically down a tube. The operator stands outside a sealed cabinet looking through a window and directs the jet of abrasive against the casting. For very large castings larger cabins are used. The operator (wearing protective clothing and breathing apparatus) enters the cabin and aims the nozzle towards the casting.

In the other type of shot blast machine the abrasive is hurled against the casting from a rapidly rotating wheel contained within the cabinet. There are many varieties of this type of machine, varying in the methods with which the castings are loaded, presented to, and moved in front of the shot impellor. Many of the common types tumble the castings and roll them together, an arrangement which is not suitable if fragile castings are being produced. In other types the castings are placed on turntables or suspended on hooks. Castings require shot-blasting for between 3 and 10 minutes, depending upon their condition.

Shot blasting machines inevitably abrade and destroy themselves, and therefore incur high maintenance costs.

Suitable abrasive, either chilled iron shot or cut steel wire or iron grit is commercially available. The use of sand for abrasive cleaning is possible, but is not recommended due to the health hazards from fine sand dust particles entering the atmosphere. In all shot blasting plant, proper dust extraction and ventilation must be arranged.

Small castings which are not too fragile may be cleaned by barrelling. The castings are loaded, together with small pieces of iron (specially made "stars", or small pieces of scrap) into a steel barrel or drum which slowly rotates so that the castings tumble and fall against each other, thus knocking off any sand, and also smoothing the metal surface.

After cleaning it is likely to be necessary to carry out further work on the casting to remove the remains of runners and risers and also to remove the evidence of the joints of the mould and the cores and any other irregularities. For castings in small quantities the finishing may be carried out with hand tools such as hammer and chisel and file. However for large castings, and large volumes of production, power tools are needed. For non-ferrous metals the use of a band saw is convenient for the removal of ingates, runners and feeders (when purchasing a band saw it is necessary to ensure that supplies of blades will continue to be available).

The joint flashes and the stubs of the ingates and other unwanted projections are removed by grinding. Many types of grinding machine are used for cleaning castings, including small hand-held grinding wheels pneumatically or electrically driven. For higher production rates floor mounted pedestal grinders are used.

The use of grinding wheels calls for good dust extraction and ventilation and proper workshop practice to avoid the dangers which may arise if they are used incorrectly. Grinding wheels are available in various specifications, according to the type of metal to be ground and the speed of the grinding machine.

If a foundry is producing castings to be machined in its own machine shop, it may be possible to economise on the grinding and cleaning operations by combining them with the machining process.

Double-Ended Grinding Machine for Cleaning Castings

CHAPTER 9

INSPECTION

The production of castings is always attended with a risk of producing substandard or defective pieces which must be rejected.

Every foundry should attempt to check its products so that the rejects are discovered and replaced within the foundry rather than submitted to the customer. Depending upon the type of casting being produced and the equipment and technical skill of the foundry, reject rates between 2% and 12% may be expected, and should be allowed for in costing and capacity predictions.

Many defects are visible on the outside of the castings. Such defects include cracks, blow holes caused by gas liberated during solidification, areas where the metal has not completely filled the mould due to low pouring temperatures, inclusions due to slag which has not been properly cleaned from the metal, and inclusions due to sand from dirty moulding or from local collapse of mould or cores.

Other defects may have to be checked in other ways. For example it may be necessary to measure the dimensions of castings which are liable to distortion or to core movement during casting. Some castings are subject to internal porosity, which is difficult to detect without extremely expensive special purpose ultrasonic or X-ray equipment. It may be useful to cut up one or two castings (perhaps castings rejected for some other reason) on a sample basis.

Much information can be gained by weighing castings. Many defects arise if the mould is not sufficiently well compacted, thus allowing the casting to swell within the mould. Such castings may weigh more than good ones.

Most castings have to be machined before they are put into final use. Machining, whether by turning, drilling, milling or other process, requires the metal to have consistent quality and not be too hard. Hardness, or "chill" (see Glossary), may occur in the production of iron castings. A simple test for this is the use of a file on the corners. With some experience it is possible to detect whether or not a casting is chilled and thus whether or not it is suitable for machining. Hardness testing machines are expensive but are essential if high quality specifications have to be met.

CHAPTER 10

SAFETY

Especially to the inexperienced observer a foundry may seem to be a dangerous place; however with sensible precautions this need not be so. Every foundry whether large or small should take safety precautions into consideration.

The most obvious source of danger is from molten metal. The following precautions are necessary:

1. Ladles and furnaces must be properly constructed and maintained and the lining should be inspected and renewed whenever necessary.
2. All furnaces, ladles, moulds and floors must be dry. Molten metal, especially cast iron, can explode if it comes into contact with water.
3. Moulds should be properly clamped and weighted together.
4. Molten metal and furnaces should always be attended by experienced and qualified people with proper supervision and equipment.
5. People working anywhere near molten metal should wear protective clothing e.g. goggles, gloves, solid boots, and other clothing as appropriate (aprons, jackets, hats, etc.) depending upon the actual job being done.

Many foundry accidents result from other causes, especially cuts and injuries from poor handling of equipment or castings.

Lifting equipment should be properly designed, be strong enough for the work and not damaged by wear or heat. Operators should be trained to use it safely.

Heavy duty leather gloves should be worn if castings or scrap with sharp corners are to be handled. Sand should be sieved to remove sharp pieces of metal, and anyone who is grinding or chipping castings should wear goggles.

Longer term health problems can be caused by breathing foundry dust especially dust from sand moulds. Operators who are knocking out, grinding, handling or cleaning sandy castings should wear a mask type respirator to keep the dust from the mouth and nose.

Every foundry should have a supply of elementary first aid materials and one or more employees should have basic first aid training.

The most important safety precaution is to ensure that employees receive proper training in the processes and use of equipment before starting to work with anything unfamiliar. External assistance may be as important from this aspect as from the technical point of view.

CHAPTER 11

FOUNDRY PLANNING

Specification

The first stage in planning a foundry is to specify the required production. The size of castings, type of castings, metals to be melted, production rate, total production, quality requirements, etc. should all be set down. Few foundries can be planned with a completely detailed accurate knowledge of all possible future developments, so sensible averages and growth factors have to be assumed. In most cases some flexibility and capacity for future changes and expansion should be included.

Using the information contained in the previous pages it will then be possible to specify the processes to be used, and to estimate and calculate the plant capacity.

A number of other factors, as considered below, should also be borne in mind.

Site and Buildings

Space

A foundry must have sufficient room for the processes to be used and the type of castings to be made. If the foundry is not mechanised and moulds are placed on the floor, space requirements might be about 10 to 15 square metres of mould pouring area per tonne production per week — although the type of mould can vary this requirement considerably. Space is also needed for melting, moulding, core making, cleaning and storage. In addition there should be room for an office, pattern shop and pattern stores, and an area for storage of raw materials and sand. A new foundry may start small but future needs for expansion on the same site should also be considered. It is not feasible to transport molten metal for long distances.

Position

Some foundry processes can produce smoke, smells, or fumes. It is therefore sensible to keep foundries away from and if possible down-wind of housing, schools, hospitals etc.

Buildings

Wooden buildings should not be used because of the danger of fire. Foundry roofs should always be water-proof. Surface drainage should be good with no fear of flooding since the dangers of mixing water and molten metal can lead to explosions.

As much head room as possible should be provided for ventilation with good natural or artificial draught, especially in hot climates.

A reliable power supply is necessary for many processes, and water supply is important. Good road access for transport of castings and raw materials should be assured.

Layout, General Principles

In foundry work handling and transport of metal, sand and moulding boxes, tools etc. occupies more effort and time than the actual production processes. In small foundries it is likely that much of the transport will be by hand and wheelbarrow, with cranes or hoists for heavy weights. The layout should be such as to minimise distances. Larger foundries may need the use of forklift trucks for internal transport.

In mechanised green sand foundries the sand transport system from storage hopper to mixer to moulding machines, and from shake-out through storage and back again, represents a major part of the investment. Conveyor belts occupy a large part of the foundry space.

It is especially important to minimise the distance for the transport of molten metal, since every few minutes of delay can lose a few degrees of temperature and mean higher fuel and melting costs. This applies whether metal is carried by hand in small ladles or by crane in larger ladles. Pouring areas should be near the melting furnaces. Raw material stocks should also be kept near the furnace area.

Patterns and inflammable chemicals should be stored away from molten metal.

Employee Facilities
Provision should be made for the foundry workers. Toilets and washrooms are needed, and many foundries provide showers, changing rooms with lockers, and a canteen or rest room.

CHAPTER 12

SOURCES OF FOUNDRY EQUIPMENT AND MATERIALS

Purpose-Built Equipment

Many industrial countries have a considerable industry of foundry plant and equipment suppliers. Although many of the manufacturers specialise in high production equipment which may not be suitable for small scale foundries, most include simpler and smaller scale plant within their range.

Not many manufacturers can supply every type of plant. However there are contracting companies which specialise in the design and supply of all the needs of new foundries — especially larger foundries.

Trade directories, which can be obtained through embassies and trade offices, list and categorise the various suppliers: see the bibliography on page 61.

An alternative approach is to obtain advice from consultants who will be able to assist in plant design and recommend one or more manufacturers for the equipment. Such consultants may be contacted through I.T.I.S. (see Preface), or through trade directories and information services.

Locally Made Equipment

If there is a reasonably good steel fabrication and machinery workshop, with some good engineering skills and proper equipment, it is possible to make many of the pieces of plant required by a small foundry. Such pieces of equipment include racks and containers, tables and benches, hoppers, moulding boxes, pattern plates, small ladles, sand driers, screens, simple crucible furnaces, etc.

Other pieces of equipment can be made in part, using purchased components for the more complex pieces, such as motors, gearboxes, cylinders, controls and burners.

Sand mixers, oil fired crucible furnaces, simple moulding machines, small cupolas, simple centrifugal casting machines, and other such machinery has been successfully made by unspecialised workshops, although it may be necessary to obtain outside advice on the design details.

Sometimes it is possible to adapt machinery which has been made for other purposes. Concrete mixers may be adapted to mix core sand, ovens of various types can be used for curing cores or drying moulds, and numerous pieces of agricultural plant can be used to handle foundry sand as easily as soil.

In general it should be remembered that foundry plant receives very rough treatment, and should therefore be very strongly and robustly designed and constructed.

Imported Raw Materials

Foundry raw materials can be obtained through the same channels as are suggested for purpose built plant and equipment; suppliers are listed in trade directories.

Many foundry raw materials such as scrap, alloys, pig iron and non-ferrous metal ingots are subject to wide price fluctuations. Informed and skilled buying can be important. In recent years oil prices and coke and gas prices have also escalated very rapidly. These factors require careful attention — and make it unwise for foundries to enter into long-term fixed price commitments for their own product.

Often transport represents a high proportion of costs, especially for sand and scrap. Many material prices are very dependant upon quantity. It can therefore be useful to buy in bulk so far as storage facilities, storage life and financial resources permit.

It may also be worthwhile to standardise on materials and to consider joint purchasing arrangements with other local foundries.

Local Materials

Local sources of scrap can be important, especially since transport costs tend to be high in relation to scrap prices. Care should be taken to guard against the purchase of contaminated scrap such as stainless steel, mild steel and alloys in cast iron from automobiles.

In most places in the world there are deposits of sand which are usable for foundry purposes. New sources of sand should be tested using binders and metals which are likely to be used in the proposed foundry. Suppliers of binders will often assist with sand evaluation.

Sand binders can be difficult to obtain except through foundry supply specialists. Sodium silicate and CO_2 gas are sometimes available for other purposes. Bentonite clay, which is sometimes used for oil drilling mud, can be used in foundries.

Many natural vegetable oils and starches can be used for making oil sand bonded cores, as described in Chapter 4.

Fuel and energy represent a major part of any foundry costs. Gas and oil prices vary throughout the world and relative economics may differ in different countries. Coke suitable for foundry cupolas is likely to have to be imported from an industrial country. Charcoal may be used in crucible furnaces.

CHAPTER 13

SKILLS AND TRAINING

It is often difficult to find sources for training and technical assistance in foundry work.

Some useful skills such as general engineering, workship practice, engineering drawing, wood and metal work, are often taught in technical schools or government training institutes but foundry practice subjects such as moulding, pattern making, foundry metallurgy etc. tend to be taught only in large centres in developed countries.

Use can be made of books and journals, but practical experience is invaluable, and for all but the simplest foundries it is essential to obtain the services, if only for a few months, of someone who has had foundry experience elsewhere.

Every advantage should be taken of learning from plant and equipment suppliers. Visits should be made to other foundries locally or abroad. Technical assistance from training programmes should be used. Government offices, United Nations or other development agencies and foreign embassies may have information about such programmes.

Consultants and foundrymen from developed countries on short-term training and advisory visits are another valuable source of information and should be used whenever possible. Location of sources for such assistance can be obtained from the I.T.I.S. (see Preface) and development agencies.

Training Times
There are so many variations of types of foundry, types of casting, quality requirements, effectiveness of teaching, and of individual aptitude that the following generalisations on training times can only be regarded as approximations for guidance. An attempt is nevertheless made to list the time which would normally be required working with or beside an experienced foundry operator or while attending a special foundry training course, in order to have sufficient ability to be able to take at least the first stages in operating a small-scale foundry.

APPROXIMATE TRAINING TIMES (FOR GUIDANCE)

OPERATION	TRAINING PERIOD
Pattern Design	6–12 months (for people with reasonable basic knowledge of technical drawing and workshop engineering)
Pattern Making (simple wood patterns)	3–6 months (assuming a reasonable basic knowledge of woodwork and use of tools)
Pattern Making (metal patterns for machine moulding and precision work)	12–36 months
Moulding by hand (simple work; plated patterns)	2–6 months
Moulding by hand (complex patterns)	6–24 months

Moulding on moulding machine	1–6 weeks	Foundry technician able to deal with sands, furnaces, basic casting methods and reasons for faulty castings	6–24 months (only for people with some basic technical education)
Shell moulding	1–5 days		
Hand coremaking	1–3 months		
Crucible furnace operation	1–2 months		
Cupola furnace operation	2–9 months		
Electric furnace operation	2–12 months		
Metal pouring	1–6 weeks.		

If it is possible to secure the services of an experienced foundryman as a consultant or as a manager at least for the initial months of the operation of a new foundry, then most of this training can be carried out as the foundry is established and commissioned, without the need for extra outside training.

SELECTED BIBLIOGRAPHY AND REFERENCES

1. **Sources of information on special purpose foundry equipment and materials**

UNIDO *Guide to Information Sources*, Number 5. Zurich, 1977.

British Foundry Equipment and Suppliers Association, (FESA)
2 Queensway,
Redhill, England.

Foundry Directory,
Standard Catalogue Information Services,
Medway Wharf Road,
Tonbridge, England.

Foundry Year Book (Annual),
Foundry Trade Journal,
2 Queensway,
Redhill, England.

American Export Register,
Thomas Import Publishing Co. Inc.,
1 Penn Plaza, 250 W. 34th Street,
New York, N.Y. 10001, U.S.A.

Trade Directories,
e.g. Kellys, Kingston, London.
Kompass, Zurich. Switzerland.

Manufacturers catalogues

2. **Textbooks and reference books on foundry practice**

Principles of Metal Casting Heine and Rosenthal, American Foundrymans Society, U.S.A., 1967.

Manual of Foundry Practice Laing and Rolfe, Chapman and Hall, London. 1948.

Broadsheets. British Cast Iron Research Association, Birmingham, England.

Introducing Iron Founding Sleman, Council of Iron Foundries Association, London, 1975.

Foundry Technology Beeley, Butterworths, London, 1972.

ASM Metals Handbook Volume 5, American Society of Metals, U.S.A., 1976.

Atlas of Castings Defects Institute of British Foundrymen, Birmingham, England.

3. **Concerning problems of small foundries in developing countries**

Iron Founding in Developing Countries Bhat and Prendergast, University of Strathclyde 1981
(provisional reference)

Establishment and Operation of Small Cast Iron Foundries:
The Indian Experience N.G. Chakrabarti, UNIDO, 1974

Iron Foundry: An Industrial Profile Intermediate Technology Publications Limited. London 1975

UNIDO Monographs on Industrial Development,
No. 4 *Engineering Industry*,
No. 5 *Iron and Steel Industry*,
UNIDO. Vienna. Austria

4. Journals and Periodicals

Modern Casting American Foundrymen's Society
Des Plaines. Illinois 60016

The Foundry Trade Journal
2 Queensway, Redhill England.

Metals & Materials Metals Society.
London England

Castings Australian Foundry Institute
18 Suttor Place,
Baulkham Hills, NSW 2153.

Indian Foundry Journal
Institute of Indian Foundrymen
Middleton Court 4/2 Middleton Street
Calcutta 7000 71

Other metallurgical and engineering journals sometimes carry papers on foundry topics.

In addition to the above English language journals, most developed and industrial countries have their own national language foundry journals. These journals can sometimes be obtained through embassies and trade delegations, as well as through libraries.

CONVERSION FACTORS

1 Tonne	: 0.984 Long tons
1 Tonne	: 1.102 Short (US) tons
1 Centimetre	: 0.394 inches
1 Kilogramme	: 2.205lbs
1 Litre	: 0.220 Imperial gallons
1 Litre	: 0.260 U.S. Gallons

1 Cubic Metre	: 35.3 Cubic feet
1 Kilogramme per cm²	: 14.22 lbs per square inch
1 Atmosphere	: 14.70 lbs per square inch

Temperature degrees C : Temperature degrees F – 32 × 5 ÷ 9

GLOSSARY

AIR-SET SAND Self hardening chemically bonded sand.

ALLOY Mixture of metals; a metal added to another metal to alter the properties.

ARC FURNACE Electric furnace heated by high voltage arc between the charge and electrodes.

BINDER Material for bonding sand grains together for mould or core making.

BLOW HOLE Casting defect caused by gas bubble in the metal.

BUSH Socket to receive pin or dowel for location.

CASTING Metal component made by pouring liquid metal into a shaped mould.

CASTING CAVITY Cavity in a mould to produce a casting.

CENTRIFUGAL CASTING Pouring metal into a spinning mould.

CHILL — 1 Metal insert in a mould to accelerate cooling locally.

CHILL — 2 Rapid solidification of part of an iron casting, giving very hard metal.

CO_2 PROCESS Hardening moulds or cores by passing carbon dioxide gas through sand bonded with sodium silicate (water glass).

CONTRACTION Extent by which a casting is smaller than the pattern from which it was made.

CORE Separately made sand piece to form internal cavities or undercut shapes in moulds.

CORE BOX Part of pattern equipment used to make cores.

CORE PRINT Part of a core for location and support in the mould; part of the mould or pattern designed to accept core prints.

CROSS JOINT Relative misplacement of two halves of casting or mould.

CRUCIBLE Heat resisting pot for metal melting.

CUPOLA Shaft furnace for melting iron with coke.

DIE Mould made of metal.

DOWEL Accurately made peg for location purposes.

DOWN SPRUE Vertical channel within mould to lead metal to the casting cavity.

DROSS	Thin layer of slag produced in liquid metal.	LOOSE FLASK	(See Snap Flask below)
DRY SAND	Dried clay-bonded sand.	LOOSE PIECE	Separate portion of pattern or corebox stripped separately from the main pattern.
FACING SAND	Specially prepared moulding sand for the surface of a mould.	MACHINING	Cutting metal accurately to size by turning, milling, drilling, grinding etc.,
FEEDER	Mass of metal moulded with a casting to compensate for metal shrinkage.	MALLEABLE IRON	Special cast iron with ductility, produced by heat-treating white iron.
FLASH	Thin piece of excess metal produced at a mould or core joint.	MIXER FILLER	Machine for continuously mixing sand with binders as it fills moulds or core boxes.
FLUX	Material added to a melt to help melt impurities into a slag.	MOULD	Form into which metal is poured to make castings.
FURNACE SILL	Level of a furnace at which materials are charged.	MOULDING BOX	Frame used to contain sand during mould making.
GREEN SAND	Sand bonded with clay and moisture.	MOULDING MACHINE	Machine for producing moulds.
HOPPER	Container for sand or another material.	MOULDING SAND	Sand mixed and prepared for making moulds.
INDUCTION FURNACE	Electric furnace with a coil which heats the charge by induction.	MOULD JOINT	Level or plane of separation of two halves of a mould.
INGATE	Position in which metal enters the mould cavity from the runner system.	NATURAL SAND	Sand suitable for moulding with its own naturally occurring clay.
INVESTMENT CASTING	Casting in refractory mould produced with a disposable (usually wax) pattern.	NODULAR IRON	S.G. Iron (See below)
		NON FERROUS	Metal other than iron or steel.
LADLE	Container for carrying and pouring molten metal.	PATTERN	Model of casting from which moulds are made.
LOAM	Stiff clay/sand mixture for large castings.	PATTERN PLATE	Plate or board upon which split patterns are mounted.

PATTERN SHOP	Workshop for making or repairing patterns.		special high strength ductile cast iron.
PIG IRON	Ingots of cast iron for melting.	SHELL MOULD	Thin mould of resin bonded sand produced from a heated pattern.
PIN	Peg used for location.	SHOT BLASTING	Cleaning with a jet of abrasive shot or grit.
POURING BASIN	Recess in the top of a mould above the down sprue.	SHRINKAGE	Reduction in volume of liquid metal as it solidifies.
PYROMETER	Instrument for measuring high temperatures.	SLAG	Non-metallic impurities separated from molten metal.
RAMMING	Packing and compacting sand around a pattern to make a mould.	SNAP FLASK	Moulding box which can be removed from a mould before pouring.
RECARBURIS-ER	Material added to molten iron to increase the carbon content.	SPLIT PATTERN	Pattern made in two pieces, one for each half of the mould.
REFRACTORY	Heat resisting material.	STRIP	Remove mould from pattern — or core from corebox.
RUNNER SYSTEM	Arrangement of channels within a mould to lead molten metal to the ingates.	SYNTHETIC SAND	Sand made suitable for moulding by the addition of clay.
ROTARY FURNACE	Oil or gas fired furnace which can revolve around a central axis.	TUYERE	Hole or nozzle through which air is blown into a furnace.
SAND MILL	Machine for mixing and preparing moulding sand.	VENT	Hole to allow gas to escape.
SAND SLINGER	Machine for packing sand into a mould from a high speed impellor.	VENT WIRE	Wire for making vents in moulds or cores.
SECTION	Metal thickness.	WHITE IRON	Cast iron with low carbon and silicon content, very hard.
SECTIONAL PATTERN	Pattern made in several pieces.	YIELD	Ratio of weight of good castings to weight of metal melted.
S.G. IRON	(Spheroidal Graphite, Nodular, or Ductile Iron) —		

63

APPENDIX

ASSISTANCE QUESTIONNAIRE

No general guide to foundry process selection can answer all the questions and problems which will be encountered by those intending to develop small scale foundries.

I.T.I.S. and the author of this book would be interested to give further advice and information, and to hear from small scale foundries about the problems which they encounter.

For this purpose the following questionnaire should be completed as far as possible, and sent to:

I.T.I.S., Myson House, Railway Terrace, Rugby, England.

A reply will be sent, giving such advice and suggestions as are appropriate.

QUESTIONNAIRE

1. THE FOUNDRY

1.1. Name of the Company:

1.2. Name of the individual to contact:

1.3. Address:

Town or City
Country
Telephone number

1.4. Whether it is a new foundry, or an established foundry.

1.5. If established please describe the existing operation and the desired development.

2. THE CASTINGS TO BE PRODUCED

2.1. Industries to be served:

2.2. Principal casting applications:

2.3. Metals to be melted:

2.4. Any specifications to be met:

2.5. The largest casting envisaged:

Description
Approximate Weight:
    Height:
    Length:
    Width:

2.6. The smallest castings envisaged:

Description
Approximate Weight:
    Height:
    Length:
    Width:

2.7. The maximum number required from any design per year:

2.8. The minimum number required from any design per year:

2.9. Any other important information about the castings to be produced:

3. PRODUCTION

3.1. Estimated total market available: Tonnes of castings per year, by major types:

3.2. Planned production in the first year:

3.3. Planned production in succeeding years:

3.4. Maximum production planned:

4. MANPOWER

4.1. Are any skilled and experienced

foundry personnel available? If so, describe the experience, jobs covered, dates, and other information.

4.2. Is industrial experience available in any of the following fields?
Machining
Steel works
Laboratory
Woodworking
Toolmaking
Maintenance engineering
Automobile engineering
Electrical engineering
Other engineering

4.3. Is unskilled labour available?

4.4. What other major industries operate in the district?

4.5. How many people will the foundry employ ultimately?

5. ACCESS TO TECHNICAL ASSISTANCE

5.1. Is there a University with a Metallurgy Department or a practical Engineering Department nearby?

5.2. Is there a Technical College nearby?

5.3. Is there a chemical analysis laboratory nearby?

5.4. Are there any other foundries operating nearby?
Describe type of foundry, and distance away, and whether co-operation can be hoped for.

6. RAW MATERIALS

6.1. Is ingot metal or pig iron of the required grade available locally?

6.2. Is scrap available locally through scrap merchants or does collection have to be organised? Estimate quantity of scrap available per year:

6.3. Price to be paid for local scrap:

6.4. Price to be paid for imported scrap:

6.5. Are there deposits of sand already proven for foundry use?
Natural clay bonded sand
Clay-free silica sand

6.6. If not are there deposits of sand which could be tested?

6.7. Are there local deposits of fire clay?

6.8. Are there local deposits of other types of clay?

6.9. Are there importers of foundry materials established in the country? Do they carry stocks of foundry materials?

6.10. Is water glass available locally? Is CO_2 gas available?

7. FUEL AND ENERGY

Please indicate which of the following are available locally, and at what price.

7.1. Fuel oil:

7.2. Diesel oil:

7.3. Coke (what type):

7.4. Coal:

7.5. Charcoal:

7.6. Electricity from mains supply

Voltage:
Frequency:
Maximum load:
Reliability and continuity of supply:
Price per KWH units consumed:
Price per KVA connected load:

7.7 Gas:
Piped natural:
Bottled Butane:
Bottled Propane:
Coal Gas:

8. LOCAL ENGINEERING INDUSTRY

Are local facilities available for the following?

8.1. Turning:

8.2. Drilling:

8.3. Milling:

8.4. Other machining:

8.5. Welding and cutting steel:

8.6. Bending and forming steel:

8.7. Sheet metal work:

8.8. Electrical engineering installation and maintenance:

8.9. Power generator maintenance:

8.10. Engineering design and drawing:

9. SITE AND BUILDINGS

9.1. Is a site available?

9.2. Are existing buildings to be used? (If so please attach a dimensioned sketch)

9.3. Are the following services available at the site?

Water
Drainage
Gas
Electrical power
Road Access
Oil Storage

10. FINANCE AND ECONOMICS

10.1. Is finance available for the project?

10.2. Is foreign exchange available for the project?

10.3. What is the maximum investment envisaged for:

Plant and equipment (imported)
Plant and equipment (local)
Stocks, and other working capital including work in progress.

10.4. What is the minimum acceptable return on investment?

10.5. What estimates have been made for the following?

Selling prices of castings, per tonne
Raw material prices, per tonne
Fuel and energy prices
Labour costs per year — unskilled
                      — skilled
Overhead costs per year (rent of premises, salaries of staff, power telephone and clerical costs, transport, local taxes, etc.)

11. TIME SCALE

11.1. When is the foundry to be built?

11.2. When is the foundry to be commissioned?

11.3. When is full production to be attained?

12. PROBLEMS

Please list any particular problems or queries with which assistance is required.

www.ingramcontent.com/pod-product-compliance
Lightning Source LLC
Chambersburg PA
CBHW080926050426
42334CB00055B/2825